Low-Frequency Sound and Marine Mammals

Current Knowledge and Research Needs

Committee on Low-Frequency Sound and Marine Mammals Ocean Studies Board
Commission on Geosciences, Environment, and Resources National Research Council

NATIONAL ACADEMY PRESS
Washington, D.C. 1994

NOTICE: The project that is the subject of this report was approved by the Governing Board of the National Research Council, whose members are drawn from the councils of the National Academy of Sciences, the National Academy of Engineering, and the Institute of Medicine. The members of the committee responsible for the report were chosen for their special competence and with regard for appropriate balance.

This report has been reviewed by a group other than the authors according to procedures approved by a Report Review Committee consisting of members of the National Academy of Sciences, the National Academy of Engineering, and the Institute of Medicine.

Support for this project was provided by the U.S. Office of Naval Research Contract No. N00014-92-J-1560/R.

Library of Congress Catalog Card No. 93-61700
International Standard Book Number 0-309-05025-1
Additional copies of this report are available from: National Academy Press 2102 Constitution Avenue, NW Box 285 Washington, DC 20055 800-624-6242 202-334-3313 (in the Washington Metropolitan Area)

B-267

Printed in the United States of America

Cover art by Ellen Hill-Godfrey. Ms. Hill-Godfrey received her Master of Fine Arts degree from the University of North Carolina-Chapel Hill. Her paintings and prints have been exhibited in the Washington, DC, area and throughout the Mid-Atlantic and Southern regions. She has done illustrations for the University of Georgia Press and the University of North Carolina's *Endeavors* magazine. She lives in Germantown, MD, and teaches at The Barnesville School.

COMMITTEE ON LOW-FREQUENCY SOUND AND MARINE MAMMALS

David M. Green, University of Florida, *Chairman*
Harry A. DeFerrari, University of Miami
Dennis McFadden, University of Texas
John S. Pearse, University of California at Santa Cruz
Arthur N. Popper, University of Maryland
W. John Richardson, LGL Ltd., King City, Ont., Canada
Sam H. Ridgway, Naval Ocean Systems Center at San Diego
Peter L.Tyack, Woods Hole Oceanographic Institution

Staff

Mary Hope Katsouros, *Director*
Robin Peuser, *Project Officer*
LaVoncyé Mallory, *Senior Secretary*
Curtis Taylor, *Office Assistant*

OCEAN STUDIES BOARD

COMMISSION ON GEOSCIENCES, ENVIRONMENT, AND RESOURCES

THE NATIONAL ACADEMIES

National Academy of Sciences
National Academy of Engineering
Institute of Medicine
National Research Council

The **National Academy of Sciences** is a private, nonprofit, self-perpetuating society of distinguished scholars engaged in scientific and engineering research, dedicated to the furtherance of science and technology and to their use for the general welfare. Upon the authority of the charter granted to it by the Congress in 1863, the Academy has a mandate that requires it to advise the federal government on scientific and technical matters. Dr. Bruce M. Alberts is president of the National Academy of Sciences.

The **National Academy of Engineering** was established in 1964, under the charter of the National Academy of Sciences, as a parallel organization of outstanding engineers. It is autonomous in its administration and in the selection of its members, sharing with the National Academy of Sciences the responsibility for advising the federal government. The National Academy of Engineering also sponsors engineering programs aimed at meeting national needs, encourages education and research, and recognizes the superior achievements of engineers. Dr. Wm. A. Wulf is president of the National Academy of Engineering.

The **Institute of Medicine** was established in 1970 by the National Academy of Sciences to secure the services of eminent members of appropriate professions in the examination of policy matters pertaining to the health of the public. The Institute acts under the responsibility given to the National Academy of Sciences by its congressional charter to be an adviser to the federal government and, upon its own initiative, to identify issues of medical care, research, and education. Dr. Harvey V. Fineberg is president of the Institutedicine.

The **National Research Council** was organized by the National Academy of Sciences in 1916 to associate the broad community of science and technology with the Academy's purposes of furthering knowledge and advising the federal government. Functioning in accordance with general policies determined by the Academy, the Council has become the principal operating agency of both the National Academy of Sciences and the National Academy of Engineering in providing services to the government, the public, and the scientific and engineering communities. The Council is administered jointly by both Academies and the Institute of Medicine. Dr. Bruce M. Alberts and Dr. Wm. A. Wulf are chair and vice chair, respectively, of the National Research Council.

www.national-academies.org

Preface

By passage of the Marine Mammal Protection Act (MMPA) of 1972, the United States acknowledged through public policy that marine mammals are a valuable national resource to be protected. Various human activities affect marine mammals and such activities are being discussed as part of the national debate during the reauthorization of the MMPA. A crucial element in the debate is the conflicting demands of protecting and conserving marine mammals versus allowing human activities that are considered important to the nation but that might affect marine mammals.

In particular, the issue of whether and how low-frequency sound might affect marine mammals has escalated to national attention because of a proposal to repeatedly measure the speed of sound in the ocean over time to determine if the ocean and the global atmosphere are warming. The Acoustic Thermometry of Ocean Climate (ATOC) project proposes to use low-frequency sound along long-distance undersea paths. This is an important issue because ocean scientists employ low-frequency sound in: geophysics research, to determine the structure of the ocean sediment and crust; ocean acoustic tomography, to study three dimensional structure of the water column over distances from 100-1000 km; ocean acoustics, to study the acoustical properties of the ocean such as propagation in ocean waves and reflections from the surface and bottom of the sea; and biological oceanography, to study marine organisms such as populations in the deep scattering layers.

In 1992, the National Research Council (NRC) established a Committee on Low-Frequency Sound and Marine Mammals under the auspices of the NRC Ocean Studies Board. The committee was chaired by Dr. David M. Green and included a range of expertise: marine mammal behavior and vocalizations, marine mammal physiology, marine mammal bioacoustics, fish bioacoustics, marine ecology, underwater acoustics, and human hearing.

This report is comprehensive and timely, contributing important information relevant to the current debate about the potential effects of low-frequency sound on marine mammals. As the public discussion intensifies, this report serves as an objective review of the current state of knowledge on this subject, recommends changes in the regulatory process to improve acquisition of scientific knowledge (not just about marine mammals but also about the ocean and Earth), and also proposes experiments that should provide the needed information to evaluate the effects of intense low-frequency sounds on a variety of marine mammals and their major prey.

WILLIAM J. MERRELL
CHAIRMAN, OCEAN STUDIES BOARD

Acknowledgments

The committee would like to acknowledge the contributions to the development of this report by several individuals who provided background information, including presentations to the committee, and who participated in several discussions on the topic of the effects of low-frequency sound and marine mammals. The committee would like to express its gratitude to Dr. Whitlow Au, Dr. Aleta Hohn, Dr. Joal Newcomb, Dr. Robert Hofman, Dr. Greg Silber, Dr. Dan Costa, Ms. Carol Fairfield, Dr. Melbourne Briscoe, Dr. Steven Swartz, Dr. Darlene Ketten, Mr. Charles Malme, Dr. Alan Berman, and Mr. Donald Baur for their valuable contributions. In addition, the committee would like to thank Mr. William Schevill, Dr. William Watkins, Ms. Nina Young, and Mr. Bruce McKay for their willingness to meet and discuss this topic with the committee.

Contents

List of Figures and Tables

Figures

Tables

Low Frequency Sound and Marine Mammals

Executive Summary

In 1992, the Committee on Low-Frequency Sound and Marine Mammals was established under the auspices of the Ocean Studies Board of the National Research Council. The committee's charges were to review the current state of knowledge and ongoing research on the effects of low-frequency [1 to 1,000 hertz (Hz)] sound on marine mammals and to advise the sponsors of the report about the effects of low-frequency sound on marine mammals. In addition, the committee was asked to consider the trade-offs between the benefits of underwater sound as a research tool and the possibility of its having harmful effects on marine mammals.

LIMITATIONS OF CURRENT KNOWLEDGE

Data on the effects of low-frequency sounds on marine mammals are scarce. Although we do have some knowledge about the behavior and reactions of certain marine mammals in response to sound, as well as about the hearing capabilities of a few species, the data are extremely limited and cannot constitute the basis for informed prediction or evaluation of the effects of intense low-frequency sounds on any marine species.

As discussed in Chapter 1 of this report, marine mammals[1] use sound to sense their environment and to communicate among themselves. Dolphins and other toothed whales use echoes from the high-frequency sound pulses they produce in order to navigate and to locate prey. They use lower-frequency sounds for communication among individuals. Baleen whales are known to produce low-frequency sounds that can propagate over long distances. Although the exact functions of most of these sounds are not yet known, they clearly play an important role in the life of the species.

The ocean is a noisy place. There are many sources of sound, and sound travels efficiently in water. Natural ocean sounds are produced by wind, waves, precipitation, cracking ice, seismic events, and marine organisms. The hearing abilities of marine mammals undoubtedly evolved to deal with these natural sounds—the ambient noise levels of the ocean. Since the advent of the industrial age, sounds made by human beings have combined with natural ocean sounds, resulting in elevated noise levels, primarily in the frequency region below 1,000 Hz. Some of the more intense human-made sounds come from ocean-going vessels, especially larger ships such as super-tankers. Other human-made sources include (1) explosive and nonexplosive seismic sounds used in geological exploration for oil and gas; (2) dredging, drilling, and marine construction; (3) surface vessels (for example, commercial, military, and pleasure boats) and submarines that use sonar to locate targets; and (4) the sound sources used by acoustical oceanographers.

The committee could find almost no quantitative information with which to assess the impact of low-frequency noise on marine mammals. For those few marine mammals on which data are available about their hearing sensitivity, it appears that low-frequency sound, even at very high levels, is barely audible to them. In addition, the range of frequencies by which these animals are affected appears to vary *among*, as well as *within*, the three different orders of Mammalia to which they belong (see Appendix B for the classification of marine mammals in the three orders). Certainly data on the hearing sensitivities of several Odontoceti (examples include the white whale, bottlenose dolphin, harbor porpoise, and false killer whale) and Pinnipedia (for example, several seals and the California sea lion) suggest that sounds below about 100 Hz are practically inaudible to these mammals (see the subsection on Acoustic Characteristics of Marine Mammal Hearing Organs, and Figure 2, in Chapter 1). But even these data are

[1] This report concentrates on cetaceans (whales, dolphins, and porpoises) and pinnipeds (seals, sea lions, and walruses) and only briefly mentions manatees, polar bears, and sea otters.

extremely limited and cannot be used to evaluate the effects of intense low-frequency sounds on all species of marine mammals.

There are literally no data available on the auditory sensitivity of any baleen whales (examples of baleens are the gray whale, right whale, fin whale, and humpback whale), although evidence from several sources suggests that members of this suborder are much more sensitive to low-frequency sounds than are members of the toothed whale suborder. The implications of this paucity of audiometric information include profound uncertainty about the interfering effect of any potential sound source on baleen whales. For example, depending on these animal's auditory sensitivity, the effect of loud low-frequency sound could conceivably range between potential hearing damage and gradual deafness for the entire species—and eventual extinction—or practically no discernible impact. Such uncertainty prevents the committee from reaching any but the most general conclusions in this regard.

There have been some observational or experimental studies and numerous anecdotal reports about the responses of marine mammals to certain sounds. Rather than summarize the existing reviews, the committee decided that its efforts could be more usefully directed to a discussion of the implications of the existing information. The committee noted, for example, that missing in most of these anecdotal accounts is information on the level of the sound exposure experienced by individual animals. Typically, neither the source level nor the received level was measured. Even when the approximate level at the source was known, the received level near the animal was usually not measured, and if it was, there were often uncertainties associated with estimating that level.

This dearth of scientific evidence makes it virtually impossible to predict the effects of low-frequency sound on marine mammals, especially on baleen whales. In the absence of such an impact assessment, the committee finds itself unable to fulfill the second part of its charge, namely, to balance the costs and the benefits of "underwater sound as a research tool" versus "the possibility of harmful effects to marine mammals." Rigorous experimental research on marine mammals and their major prey is required to resolve the issue of how low-frequency sound affects these species. The committee recommends that future experiments be conducted in such a manner that the received level of the sound and the behavior of the animal can be studied together. Such investigations may be logistically complex and may require permits, which are sometimes difficult to obtain. Chapter 2 discusses the permit issue, and Chapter 3 describes the types of studies needed.

CHANGES PROPOSED IN REGULATORY STRUCTURE

It is the committee's judgment that the regulatory system governing marine mammal "taking" by research actively discourages and delays the acquisition of scientific knowledge that would benefit conservation of marine mammals, their food sources, and their ecosystems. The committee thus proposes several alternatives for reducing unnecessary regulatory barriers and facilitating valuable research while maintaining all necessary protection for marine mammals.

Although the committee strongly agrees with and supports the objective of marine mammal conservation, it believes that the present regulation of research is unnecessarily cumbersome and restrictive.[2] Not only is research hampered, but the process of training and employing scientists with suitable research skills is impeded by this system. Better and more humane management of marine mammals depends on understanding them better. Well-trained researchers are the ultimate source of our knowledge about marine mammals. The present system, in effect, impedes acquisition of the information and understanding needed to pursue a more effective conservation policy.

At present, one of two types of regulatory approval is necessary to pursue any scientific research that might "take" marine mammals. "Take" is defined in the Marine Mammal Protection Act of 1972 as harass, hunt, capture, or kill. The two types of approval are (1) a scientific research permit, which currently is available only for research *on or benefiting marine mammals*, or (2) a small incidental take (SIT) authorization.[3] Harassment has been interpreted as any action "that results in an observable change in the behavior of a marine mammal." This interpretation of harassment seems to the committee to be inappropriately broad. Although a scientific research permit may be granted a few months after application, the SIT authorization can take up to two years and may require public hearings. Any general biological investigations of nonmarine mammals, as well as

---

[2] Under the Marine Mammal Protection Act of 1972 and the Endangered Species Act of 1973, the National Marine Fisheries Service of the U.S. Department of Commerce has responsibility for all cetaceans and all pinnipeds except walruses. The U.S. Fish and Wildlife Service in the Department of the Interior has responsibility for walruses, sea otters, manatees, and polar bears.

[3] A small incidental take authorization refers to the provision in the Marine Mammal Protection Act of 1972 (Sect. 101(a)(5)) that allows for the incidental, but not intentional, taking of a small number of marine mammals within a specified geographical region by U.S. citizens engaged in a specified activity other than commercial fishing.

general physical investigations of the ocean and planet, when carried out in the ocean, may disturb marine mammals to some degree. Such investigations presently require, at least in theory, SIT authorizations. Neither general biological nor physical research is eligible for the scientific research permit, unless it can be shown that the investigations might benefit conservation of marine mammals by providing useful information concerning their habitats or the food chains.

The committee compared the effects of scientific research on marine mammals with the effects from commercial fishing and shipping. Commercial fisheries have had a blanket exemption[4] under the Marine Mammal Protection Act of 1972 from the prohibition on taking marine mammals, and can kill marine mammals even from depleted populations. Furthermore, the noise from passing marine traffic, including supertankers, is not regulated as harassment or as a "take." Scientific research, especially oceanographic surveys, seldom results in any fatalities among marine mammals, although the sound levels produced in these investigations are intense and could potentially affect large populations. The shipping industry, especially super-tankers, produces much more total noise than scientific research does, if measured on a yearly basis. Making an exact comparison is difficult, because the dominant sound frequencies produced by super-tankers are much lower than those produced by oceanographic experiments, and the hearing sensitivity of the affected animals is not known. Given these facts, it is illogical to regulate scientific sources, some of which have the potential to produce answers about the impact of noise on marine mammals, while ignoring other significant sound sources that are far more common, such as supertankers. To regulate intelligently, we need more information about the auditory systems of marine mammals and the impact of all noise sources.

The committee considered several possible alternatives for facilitating valuable research while maintaining all necessary protection for marine mammals. One alternative would be to incorporate scientific researchers as "other users" in the regulatory regime recently proposed by the National Marine Fisheries Service (NMFS) of the U.S. Department of Commerce to govern commercial fishing and marine mammal interactions. Another alternative would be to establish a decentralized regulatory regime, possibly patterned after the Institutional Animal Care and Use Committee (IACUC) system currently used to monitor research conducted on nonmarine animals in scientific laboratories.

[4] This exemption under the Marine Mammal Protection Act of 1972 expires May 1, 1994.

If the existing system of regulations is maintained, the committee urges that steps be taken to expedite the small incidental take authorization process for all scientific activities involving nonlethal takes, and to further simplify the process for nonlethal takes producing only negligible impact. The committee suggests rewording those provisions to delete references to effects on "small" numbers of marine mammals, provided that the effects are negligible. It would also be beneficial to broaden the definition of research for which scientific permits can be issued to include activities beyond those directly "on or benefiting marine mammals."[5] In order to place regulations on a more rational footing, the population status of each species should determine the number and types of allowable takes, and the same regulations should apply equally to all activities, scientific and otherwise. The committee notes that some of these recommendations would require congressional action to change the Marine Mammal Protection Act and perhaps other laws. However, other recommendations could be implemented under existing laws through changes in regulations.

The committee is by no means recommending that scientific research be afforded a blanket waiver of the provisions of the Marine Mammal Protection Act, the Endangered Species Act, and the National Environmental Policy Act that relate to taking of marine mammals, either during research on marine mammals or on other topics where experiments might incidentally affect marine mammals. Rather, the committee urges a more logical balance between the regulation of research and other human activities, and a more expeditious permitting process. Appropriate scientific research might identify the sources of human-made noise that actually endanger marine mammals, and may suggest regulation of certain sound sources that are presently unregulated. This research could provide information that would benefit all marine mammals.

Finally, the committee considered the "120-decibel (dB) criterion" that is regarded in some contexts as a noise level above which potentially harmful acoustic effects on marine mammals might occur. In the opinion of the committee, the data from which the 120-dB criterion was derived are being overly extrapolated, largely because of the scarcity of experiments providing quantitative information about the behavior of marine mammals in relation to sound exposure. It is possible that this level is simply the one at which the animals de-

[5] It is the committee's view that the recent proposed rule changes make it clear that NMFS intends to make regulations on general scientific research even more restrictive (58 Fed. Reg. 53,320–53,364 (1993)).

tected the presence of a sound. If this is true, then there is no scientific evidence to indicate that the relatively minor and short-term behavioral reactions observed indicate any significant or long-term effects on the animals. Marine mammals, like other animals, respond to many stimuli, natural and human-made. These reactions are part of their normal behavioral repertoire and are not necessarily indicative of an adverse effect.

One danger of adopting a single number, as with the 120-dB criterion, is in applying it to all species of marine mammals and to all sounds and situations, regardless of the frequency spectrum, regardless of the temporal pattern of the sound, and regardless of differences in the auditory sensitivity of the different groups of marine mammals. There is general agreement that these variables are important in determining whether the 120-dB figure is appropriate in any given situation.

RECOMMENDED RESEARCH

The research that would provide some of the missing information is conceptually straightforward biological science, the proposed experiments should provide much of the needed information, and the cost is not enormous compared with that of other scientific efforts of comparable magnitude.

The committee's aim was to identify general research needs that are crucial to a full evaluation of the effects of intense low-frequency sounds on a variety of marine mammals and their major prey. The committee has identified the following general areas in which more information must be developed:

1. Research on the behavior of marine mammals in the wild.
2. Research on the auditory systems of marine mammals.
3. Research on the effects of low-frequency sound on the food chain of marine mammals.
4. Development and application of measurement techniques to enhance observation and data gathering.

Chapter 3 presents the committee's recommendations for specific types of studies that are needed in each of these general areas.

The committee recommends that an accelerated program of scientific studies of the acoustic effects on marine mammals and their major prey (including the studies described in Chapter 3) be undertaken. These studies should be designed to provide information needed to direct policies that will provide long-term protection to the species.

1

A Review of Current Knowledge

In 1992, the Committee on Low-Frequency Sound and Marine Mammals was established under the auspices of the Ocean Studies Board of the National Research Council. The committee was charged with (1) reviewing the current state of knowledge and ongoing research on the effects of low-frequency [1 to 1,000 hertz (Hz)] sound on marine mammals, (2) advising the sponsors about the effects of low-frequency sound on marine mammals, and (3) considering the tradeoffs between the benefits of underwater sound as a research tool and the possibility of its having harmful effects to marine mammals.

The report is organized as follows. Chapter 1 discusses current knowledge with respect to changing levels of sound in the ocean and some possible effects on marine mammals. It then describes the state of knowledge regarding components that must be determined to evaluate a potential source of noise interference affecting marine mammals. The committee then discusses the gaps in knowledge, including those on auditory sensitivity and behavioral responsiveness of marine mammals, that would require research before predictions can be made of the effects of low-frequency sound on marine mammals. Chapter 1 closes with a discussion of the difficulties related to a sound criterion that is being applied to underwater noise.

Citing an urgent need for additional research, the committee takes up the question of the regulation of research on marine mammals and related topics under the Marine Mammal Protection Act of 1972.

Chapter 2 reviews the permitting process as it now stands, and presents the committee's recommendations for change in the regulatory structure in order to facilitate needed research.

Chapter 3 outlines the research that would provide some of the missing information needed for a better understanding of the effects of low-frequency sound on marine mammals and their prey. The chapter is divided into sections on Behavior of Marine Mammals in the Wild, Structure and Function of the Auditory System, Effects of Low-frequency Sounds on the Food Chain, and Development and Application of Measurement Techniques.

Appendix B of this report provides an introduction to marine mammals, including tables listing the species by their scientific classification. The tables indicate the species for which audiometric information is available.

CHANGES IN OCEAN SOUNDS AND POTENTIAL EFFECTS ON MARINE MAMMALS

Noise is widely acknowledged to be an environmental pollutant for humans and many other terrestrial species, and it is no doubt a pollutant for marine animals as well. All animals, whether they are terrestrial or aquatic, have evolved to live in particular ecological niches. The physical characteristics of these niches have established various constraints on the perceptual systems and communication abilities of these organisms. Consequently, the hearing abilities of marine mammals[1] undoubtedly evolved to deal with the ambient noise levels in the aquatic environment prior to the industrial age.

Terrestrial mammals depend on sound to analyze and interact with their environment and to communicate among themselves. Marine mammals, including cetaceans (various whales, porpoises, and dolphins) and pinnipeds (seals, sea lions, and walruses), are probably even more dependent on sound and hearing than terrestrial mammals are. For example, it is known that dolphins use the echoes from the high-frequency sounds they produce in order to navigate and to locate prey. They use moderate-frequency sounds to maintain contact among individuals. Baleen whales are known to produce low-frequency sounds that propagate over long distances. Although the

[1] This report concentrates on cetaceans (whales, dolphins, and porpoises) and pinnipeds (seals, sea lions, and walruses) and only briefly mentions manatees, polar bears, and sea otters.

exact functions of most of these sounds are not yet known, they clearly play an important role in the life of the species.

The natural sounds of the ocean—produced by wind, waves, precipitation, natural seismic events, and creatures that live in the sea—create a noisy environment (Wenz, 1962; Tavolga, 1967). There are many sources of sound, and sound travels efficiently in water. In recent times, the background sound level has increased considerably because humans have developed a number of highly intense sound sources (Ross, 1976; Urick, 1983, 1986). Underwater sound produced by numerous ocean-going vessels is one of the more intense human-made intrusions (see Figure 1). Supertankers produce spectrum levels that equal or exceed 200 decibels (dB) [reference (re) 1 micropascal (μPa) at 1 meter (m) —water standard[2]] in the very low (<10 Hz) region of the spectrum (Cybulski, 1977; Leggat et al., 1981). Explosive and nonexplosive seismic sounds used in geological exploration for oil and gas, as well as sounds from dredging, drilling, and marine construction, also contribute to the sonic burden of the sea (Richardson et al., 1991). Surface vessels and submarines employ active sonar, an apparatus that uses sonic or ultrasonic waves to locate submerged objects and in the process introduce brief, high-intensity pulses that sometimes propagate great distances. Acoustic oceanographers use intense sounds, especially low-frequency sounds, to study the physical properties of the ocean (Spindel and Worcester, 1990; Baggeroer and Munk, 1992; Worcester et al., 1993). Such human-made sounds combine with the natural sounds of the ocean, and elevate the ambient noise level, mainly in the frequency region below 1,000 Hz (Figure 1).

Major changes in human-made noise, such as the change resulting from the advent of supertankers, can be assumed to affect the ability of marine mammals to communicate and to receive information about their environment. Such noise may interfere with or mask the sounds used and produced by these animals and thereby interfere with their natural behavior.

Higher levels of human-made sounds can cause obvious disruptions: they may frighten, annoy, or distract the animals and lead to physiological and behavioral disturbances. They can cause reactions that might include disruption of marine mammals' normal activities

[2] The reader should be aware that different reference pressures are used to specify sound levels in air and in water (see Appendix A for additional information). When decibel numbers are given to specify sound levels in the text of this report, the reference number, together with the designation "water standard" or "air standard," is also given.

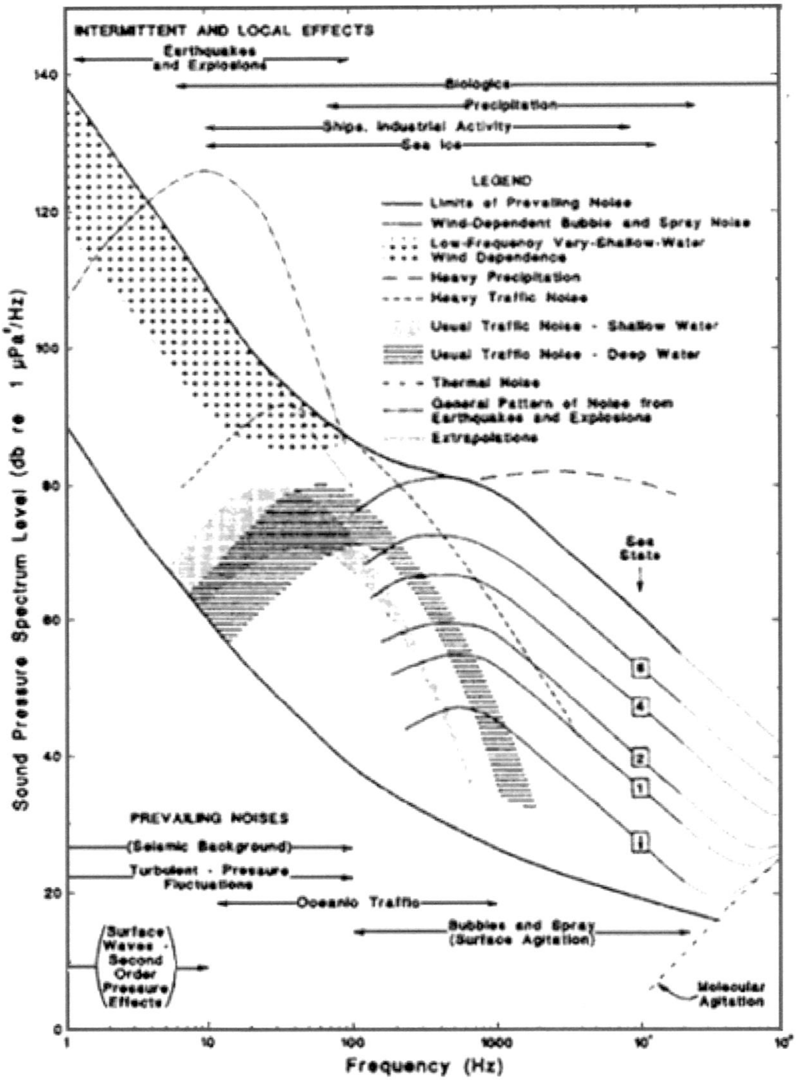

Figure 1
Ambient noise spectra. Source: Adapted from Wenz (1962), courtesy of LGL Ltd. NOTES:
The left ordinate uses the water standard for decibel calculation. The values in this figure are averages over long periodsof time; the variability in the levels is quite large.

and, in some cases, short-or long-term displacement from areas important for feeding or reproduction. They may also disturb the species such as fishes, squids, and crustaceans upon which the marine mammals prey.

At still higher levels, human-made noise may cause temporary or permanent hearing impairment in marine mammals. Such impairment would have the potential to diminish the individual's chances for survival.

For some years, there has been growing general concern about the potential effects of human-made underwater sounds on marine mammal communication (e.g., Payne and Webb, 1971; Reeves, 1977; Myrberg, 1978; Acoustical Society of America, 1981). During the 1980s, a number of more specific studies were undertaken to determine the reactions of several cetacean species and, to a lesser degree, pinnipeds to noise from oil industry operations, shipping, tourist activities, and rocket launches (reviewed in Richardson et al., 1991). Recently, concern has also arisen about the potential effects of strong low-frequency sounds used by acoustic oceanographers, underwater acousticians, and operational naval activities (Mulroy, 1991; Simmonds and Lopez-Jurado, 1991).

EVALUATING NOISE INTERFERENCE

For a sound signal to be detected by a marine mammal or its prey, the signal must contain enough energy to exceed the ambient (background) level of sounds near the animal at frequencies near the frequency of the signal. This effective signal-to-noise ratio is determined by four components: (1) the source level of the signal; (2) the transmission loss in the sea—the decrease in intensity as sound travels to the animal; (3) the level of other sounds near the animal (ambient noise level); and (4) the auditory characteristics of the receiver, the hearing organ. Any potential source of noise interference must be evaluated by determining these four components of the source-path-receiver model. Specific information is often known or can be obtained more easily about the first component, the source level of the signal, in comparison to the other three components, which are more technically complex. The other three components are discussed briefly below.

Transmission Loss of Sound in the Ocean

The second component, the transmission loss of sound in the ocean, is a complicated matter but one that is reasonably well understood, both in theory and in practice (Urick, 1982, 1983). Studying the transmission of sound in the sea has been a major research area of modern oceanography and of the world's naval forces. The transmission loss problem is complicated by the presence of gradients of temperature, pressure, and salinity (gradients that cause variation in the sound speed of the medium) and by the variable effects of the sea bottom and surface on sound transmission. These phenomena result in refractions and reflections of sound waves. Despite these many factors, reasonable approximations exist that permit prediction of sound levels in many underwater situations.

Ambient Noise Level of the Sea

The third component of the source-path-receiver model, the ambient noise level of the sea, has also been extensively studied (see Figure 1 and Urick, 1986). Ambient noise levels have been of considerable interest both to naval forces and to acoustic oceanographers. These levels and the details of their spectra and variation over time establish the conditions for detecting extraneous signals either by a physical receiver or by any biological organism. Extensive data and models concerning both acoustic propagation and the ambient noise level can be used to help evaluate the potential impact on marine mammals of sound from any human-made source.

Acoustic Characteristics of Marine Mammal Hearing Organs

The component of the source-path-receiver model about which the least is known is the fourth component—the acoustic characteristics of the receiver, the hearing organ. At present, very little is known about the detection and interpretation of low-frequency sounds by the animals that appear to use such signals. To understand the significance of this lack of knowledge, it is useful to think of a sensory system—in this case, hearing—as a window that allows the brain to gather information about a particular aspect of the physical environment. That is, the sense of hearing in any animal is sensitive to a limited range of sound frequencies, and outside this range the animal is unaffected by any stimulus, no matter how intense. With respect to the auditory sensitivities of marine mammals, the range of frequencies by which these animals are affected appears to vary *among*, as well as *within* , the three different orders of Mammalia to which they belong (see Appendix B for the classification of marine mammals in the three orders). As described in more detail below, the order Cetacea includes dolphins and the smaller toothed whales, whose auditory sensitivities are greatest at very high frequencies. Available data for all toothed whales that have been measured for auditory sensitivity are reasonably consistent. However, for another group of cetaceans whose hearing is undoubtedly tuned to low frequencies—the baleen whales—there are virtually *no quantitative data* on auditory sensitivity.

Audiograms have been obtained for several dolphins and smaller toothed whales. Sensitivity of hearing is best at about 20,000 Hz, with about 80 dB less sensitivity at 100 Hz for two species—the beluga whale and bottlenose dolphin —for which data are available (see Figure 2A). For reasons described below, the auditory sensitivities of these smaller cetaceans are probably quite different from those of the baleen or larger toothed whales, whose auditory sensitivities have never been assessed quantitatively. Also, even for the relatively well-studied toothed whales, audiograms have been obtained for only a few species, and in most cases for only one or two individuals per species. However, different individuals of a single species can have quite different hearing capabilities (*see*, e.g., Hall and Johnson, 1971; vs. Bain et al., 1993; Ridgway and Carder, 1993).

It is thought that baleen whales hear very low-frequency sounds because their vocalizations contain considerable low-frequency energy. Some of the dominant calls of the blue and fin whales are in the frequency range between 12 and 25 Hz, and their *source* level is about 177 dB re 1 μPa—water standard. The anatomical nature of baleen whale hearing organs also suggests that they are sensitive to low-frequency sounds (Fleischer, 1976; Ketten, 1991). Certainly there are obvious anatomical differences between the hearing organs of the baleen and the toothed whales (Ketten, 1991).

The strongest quantitative evidence of the auditory sensitivity of baleen whales comes from field observations in which human-made sounds in the low-frequency regions caused two species of baleen whales to alter their migration paths or to avoid certain areas (Malme et al., 1983, 1984, 1988; Richardson et al., 1986, 1990; Richardson and Malme, 1993; Dahlheim and Ljungblad, 1990). Recordings from a variety of sources were used in these studies, including continuous sounds such as those from oil drilling and production platforms, as well as impulsive sounds from air guns. The broadband levels of low-frequency sounds causing avoidance by about 50 percent of the gray and bowhead whales in the area of the observations occurred when the *received* levels were around 115 to 120 dB re 1 μa—water standard for the continuous sounds and about 160 to 170 dB for the pulsed sounds. These received levels are overall levels, and the sources produced energy over a few octaves. The peaks of the broadband

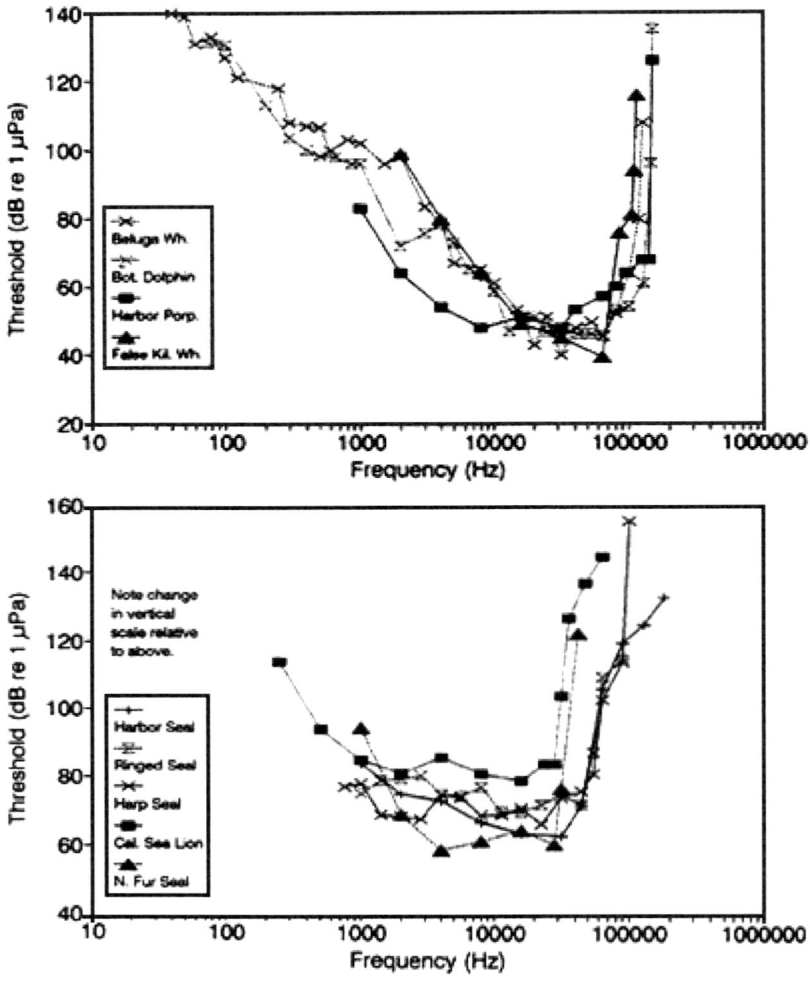

Figure 2
Underwater audiograms of selected toothed whales (A) and pinnipeds (B). Source: Adapted from Richardson et al. (1991): (A) beluga (White et al., 1978; Awbrey et al., 1988; and Johnson et al., 1989); bottlenose dolphin (Johnson, 1968); harbor porpoise (Anderson, 1970); false killer whale (Thomas et al., 1988); (B) harbor seal (Møhl, 1968); ringed seal (Terhune and Ronald, 1975); harp seal (Terhune and Ronald, 1972); California sea lion (Schusterman et al., 1972); northern fur seal (Moore and Schusterman, 1987), courtesy of LGL Ltd.

spectra were in the general region of 100 Hz. The hearing sensitivities of baleen whales are not known over any specific frequency region but are clearly different from those of toothed whales. In the low-frequency region, it is uncertain that the toothed whale could even detect continuous sounds with received levels near 115 dB re 1 μPa—water standard. The implications of this paucity of audiometric information include profound uncertainty about the interfering effect of any potential sound source on baleen whales. For example, depending on this animal's auditory sensitivity, the effect of loud low-frequency sound could conceivably range between potential hearing damage and gradual deafness for the entire species—and eventual extinction—or practically no discernible impact. Such uncertainty prevents the committee from reaching any but the most general conclusions in this regard.

Among pinnipeds, which include, for example, seals, sea lions, and walruses, underwater audiograms are also available for several species, but there are very few data on underwater hearing sensitivity at frequencies below 1,000 Hz (see Figure 2B). Based on the slopes of the audiograms at the lowest frequencies measured, it appears that some pinnipeds may have better hearing sensitivity at low frequencies than do the two odontocetes, the beluga whale and bottle-nose dolphin, for which low-frequency audiograms are available. There are no published data for elephant seals and few for walruses (Kastelein et al., in press). Elephant seals are known to dive deeply and are of special interest because they may dive deep enough to reach the sound channel [3] and thus be exposed to higher levels of sound (in Chapter 3, see the subsection on Replication of Data in the section on Structure and Function of the Auditory System). There are essentially no data about the behavioral reactions of pinnipeds as a function of the received level of human-made underwater sounds (Richardson et al., 1991). More is known about the low-frequency hearing abilities of some pinnipeds in air and about behavioral responses of certain pinnipeds to airborne sounds. The in-air information, however, cannot be applied directly to the underwater situation.

Although there has been considerable work on the anatomy of the auditory apparatus of manatees (e.g., Fischer, 1988; Ketten et al., 1992) and several relevant studies are under way, there is very little information about the hearing abilities of marine mammals in the

[3] SOFAR (Sound Fixing and Ranging) channel, also known as the deep sound channel, is a layer of seawater extending from about 700 meters down to about 1,500 meters, in which sound travels at about 1,450 meters per second, the slowest it can travel in seawater (*McGraw-Hill Dictionary of Scientific and Technical Terms*).

order Sirenia, which includes manatees and dugongs. The evoked-potential data that are available for the manatee do not extend below 1,000 Hz (Bullock et al., 1980, 1982; Popov and Supin, 1990). Recent behavioral tests indicate some detection of sound as low as 15 Hz and as high as 4 to 6 kilohertz (kHz), with the greatest sensitivity in the 6–20 kHz region (Gerstein et al., 1993). Sirenian vocalizations are known to be in the 1–8 Khz range (Nishiwaki and Marsh, 1985; Hartman, 1979). There is little information on the behavioral responsiveness of sirenians in relation to underwater sound level, although work on this topic has also begun (Weigle et al., 1993).

The committee is not aware of any specific data on the hearing sensitivity of sea otters or polar bears, both of which are considered marine mammals. There is little information on the reactions of these mammals to underwater sound (see, however, Riedman, 1984).

The scarcity of information about the auditory sensitivity of marine mammal hearing is coupled with equal uncertainty on the structural and behavioral levels. With respect to structure, for example, how does sound actually get to the inner ears of these animals? Are the pathways different for different marine mammal groups, and are they different in a given species for different sound frequencies? In terms of behavior, if human-made sounds are interfering with marine mammals, how do the animals respond to the sounds, and do they habituate when the sound is repeated? Much greater knowledge is also needed about the role of acoustical information and hearing in the behavior of such mammals in their natural (undisturbed) habitats. Only with such baseline information can it be determined that a response caused by an intrusive sound is normal.

THE NEED FOR ADDITIONAL RESEARCH

There have been some observational or experimental studies and numerous anecdotal reports about the responses of marine mammals to sound. Extensive reviews and evaluations of this literature can be found in Richardson et al. (1991) and Reeves (1992). Missing from most of these studies, however, is information on the *level* of the sound exposure: *typically neither the source level nor the received level was measured.* Even in cases where the level of the sound at the source was known, there were often uncertainties in calculating the received level near the animal. Direct measurements or reliable estimates of received levels of sound are difficult or impossible to obtain during many observational studies and usually were not obtained even when it might have been possible. There is need for planned experiments in which the received level of the sound and the behavior of the animal can be studied together. Such investigations would probably be logistically complex and would require scientific permits. Chapter 2 discusses the permit issue, and Chapter 3 describes the kinds of studies needed.

As discussed in the preceding section, the lack of knowledge about the auditory sensitivity and behavioral responsiveness of marine mammals makes it impossible to predict how low-frequency sound may affect them. The lack of such knowledge, which must serve as the basis for identifying and determining the effects of particular underwater sounds on particular species, precludes the committee from being able to obtain a reasonable estimate of costs on which to base advice about the potential trade-offs between "the benefits of underwater sound as a research tool" and "the possibility of harmful effects to marine mammals." Research on marine mammals and their major prey is urgently needed to permit an adequate response to this question. The most immediate research needs are outlined in Chapter 3 of this report.

It is the belief of this committee that an accelerated program of scientific studies of the acoustic effects of low-frequency sound on marine mammals and their prey (including the studies described in Chapter 3) should be undertaken. These studies should be designed to provide the information needed to direct policies that will provide long-term protection to the species.

ORIGINS AND DRAWBACKS OF THE "120-DB CRITERION"

The phrase "120-dB criterion" refers to a level of sound that has been identified informally as a level above which acoustic effects on marine mammals might occur. In reviewing plans for activities that produce underwater noise, it appears to the committee that the National Marine Fisheries Service of the U.S. Department of Commerce considers that marine mammals exposed to broadband received levels above 120 dB (re 1 μPa—water standard) might be affected by the sounds. [4] Almost all sound sources used in acoustical oceanography or for operational Navy purposes have a source level far above 120 dB (re 1 μPa at 1 m —water standard), as do some motorized boats. Depending on the source level, frequency, and local propagation condi

[4] Under the Marine Mammal Protection Act of 1972 and the Endangered Species Act of 1973, the National Marine Fisheries Service of the U.S. Department of Commerce has responsibility for all cetaceans and all pinnipeds except walruses. The Fish and Wildlife Service in the U.S. Department of the Interior has responsibility for walruses, sea otters, manatees, and polar bears.

tions, the received level of sound from these three sources will exceed 120 dB within a distance that could range from a few meters or tens of meters for a weak source up to many kilometers for a stronger one. Thus, if the 120-dB criterion were to be applied consistently, almost any source of human-made underwater noise—including every powerboat—might be subject to regulatory scrutiny.

The 120-dB criterion arises primarily from two series of field studies. One series involved gray whales migrating along the coast of California and summering (spending the summer season) in the Bering Sea (Malme et al., 1983, 1984, 1988). The other was a series of studies of migrating and summering bowhead whales in the Beaufort Sea (Richardson et al., 1985, 1986, 1990; Ljungblad et al., 1988; Richardson and Malme, 1993). Unlike most other field studies on marine mammals, these two series of studies provided estimates of the sound exposure level in the vicinity of the animals while their behavior was being observed. Both series demonstrated that a variety of broadband continuous sound stimuli with spectra peaking in the frequency region of 100 to 300 Hz caused a detectable change in the behavior of some animals. The received level of continuous sound that caused a reaction in about half the animals was about 120 dB (re 1 μPa—water standard). There was considerable variation, however, with some animals reacting at lower levels and some not reacting at considerably higher levels.

The 120-dB figure has been applied, at times, to other types of sound and to other species of marine mammals without regard to the frequency spectrum or temporal pattern of the sound or to differences in the auditory sensitivity of the different groups of marine mammals. These variables are undoubtedly important in determining whether the 120-dB figure is appropriate for any given situation. For example, the temporal pattern of exposure was found to be very important for both the gray whale and the bowhead whale. The average pulse pressure level of a series of brief impulses had to be 30 to 50 dB more intense (150 to 170 dB re 1 μPa—water standard) to produce the same change in the animal's behavior as did a steady-state or continuous sound at 120 dB (Malme et al., 1984; Richardson et al., 1986; Ljungblad et al., 1988). In contrast, bowhead whales sometimes react to increasing noise levels from an approaching boat when the broadband level is well below 120 dB (Miles et al., 1987:225*ff*; Richardson and Malme, 1993). Thus, the actual threshold of reaction can range from well below to well above 120 dB, depending on circumstances.

As is true of most field observations, many different interpretations of these results can be offered. For example, according to the studies mentioned, the change in behavior of the migrating gray whales was minor and brief, involving a slight deflection in the migratory path. One can argue that the animals simply detected a potential obstruction and made a relatively mild deflection in their course to avoid the obstacle. Certainly the energy expended in their response was minimal. Energetic effects were obviously greater for migrating bowhead whales. They apparently avoided an icebreaker-supported drillship by 10 to 30 kilometers (km) (LGL and Greeneridge, 1987; Brewer et al., 1993). Additionally, in the case of the withdrawal of bowheads from feeding areas, the action was observed when a novel stimulus was introduced (Richardson et al., 1990; Richardson and Malme, 1993). Such withdrawal behavior might or might not quickly habituate if the sound were repeated, but that study was not able to obtain information about habituation. Furthermore, only two species of whales were involved, and the results of the same experiment would very likely be different for other species. Because of their apparent lack of sensitivity at these low frequencies, some toothed whales, for example, may not detect sound at the levels that affected the gray whales and bowhead whales (see Figure 2A).

That the 120-dB number is considered to be such an important regulatory criterion is testimony only to the paucity of our knowledge about marine mammals.

In trying to protect human beings from the harmful effects of noise, the United States has adopted no absolute standard. It is known at what levels sound elicits the response of "highly annoyed" on questionnaires, but there are no national standards on an upper limit to such exposure. Noise is recognized as a source of stress, and we know that physiological changes can occur when subjects are exposed to certain noise levels. There is little consensus as to either the level of noise exposure that produces either harmful extra-auditory effects in humans or the dose-response relation over the long term. The one national U.S. standard with respect to human exposure to noise is the 90-dBA level (re 20 μPa —air standard) adopted by the U.S. Department of Labor's Occupational Safety and Health Administration (OSHA). It establishes that in the workplace if the exposure exceeds 90 DBA (air standard), then some kind of hearing protection program must be initiated. The 90-DBA level is about 100 dB above the level at which such a sound might first be detectable. If we assume the same dynamic range (100 dB) for whales as for humans and that 120 dB represents a detection level, then the upper limit for marine mammals might be set at 220 dB re 1 μPa (water standard).

In humans, the upper limit in the OSHA standard was determined largely on the basis of extensive scientific experiments involving induced temporary hearing loss. Thus, for humans, temporary threshold shifts provided the basic data for the adoption of the primary noise standard in the United States. The limits were subsequently confirmed in epidemiological studies of permanent hearing loss. No data on temporary threshold shifts (that is, the occurrence of temporary hearing loss) in marine mammals have been published to date.

Human annoyance to human made noise is widely variable, depending on the individual, the situation, and the characteristics of the noise. People are frequently annoyed by noise at levels well below those that introduce temporary or permanent hearing loss. The significance of human annoyance, however real, has proven difficult to evaluate. The biological significance of disturbance reactions by marine mammals will probably be even harder to determine.

REFERENCES

Acoustical Society of America. 1981. San Diego workshop on the interaction between man-made noise and vibration and arctic marine wildlife. Rep. from Acoust. Soc. Am. for Alaska Eskimo Whaling Comm., Barrow, AK. 84 pp.

Andersen, S. 1970. Auditory sensitivity of the harbour porpoise *Phocoena phocoena*. Invest. Cetacea 2:255–259.

Awbrey, F.T., J.A. Thomas, and R.A. Kastelein. 1988. Low-frequency underwater hearing sensitivity in belugas, *Delphinapterus leucas*. J. Acoust. Soc. Am. 84(6):2273–2275.

Baggeroer, A., and W. Munk. 1992. The Heard Island Feasibility Test. Phys. Today 45(9):22–30.

Bain, D.E., B. Kriete, and M.E. Dahlheim. 1993. Hearing abilities of killer whales (*Orcinus orca*). J. Acoust. Soc. Am. 94(3, Part 2):1829.

Brewer, K.D., M.L. Gallagher, P.R. Regos, P.E. Isert, and J.D. Hall. 1993. ARCO Alaska Inc. Kuvlum #1 exploration prospect/ Site specific monitoring program final report. Rep. from Coastal and Offshore Pacific Corp., Walnut Creek, CA, for ARCO Alaska Inc. and the Nat. Mar. Fish. Serv., Anchorage AK. 80 pp.

Bullock, T.H., D.P. Domning, and R.C. Best. 1980. Evoked brain potentials demonstrate hearing in a manatee (*Trichechus inunguis*). J. Mammal. 61(1):130–133.

Bullock, T.H., T.J. O'Shea, and M.C. McClune. 1982. Auditory evoked potentials in the West Indian manatee (Sirenia: *Trichechus manatus*). J. Comp. Physiol. 148A(4):547–554.

Cybulski, J. 1977. Probable origin of measured supertanker radiated noise spectra. *In*: Oceans 77 Conference Record, Inst. Electrical and Electronic Eng., New York, NY. pp. 15C-1 to 15C-8.

Dahlheim, M.E., and D.K. Ljungblad. 1990. Preliminary hearing study on gray whales (*Eschrichtius robustus*) in the field. *In*: J.A. Thomas and R.A. Kastelein (eds.), Sensory Abilities of Cetaceans/Laboratory and Field Evidence. Plenum, New York. pp. 335–346.

Fischer, M.S. 1988. Zur Anatomie des Gehörorganes der Seekuh (*Trichechus manatus L.*), (Mammalia: Sirenia). Z. Söugetierk. 53:365–379.

Fleischer, G. 1976. Hearing in extinct cetaceans as determined by cochlear structure. J. Paleontol. 50 (1):133–152.

Gerstein, E.R., L.A. Gerstein, S.E. Forsythe, and J.E. Blue. 1993. Underwater audiogram of a West Indian manatee (*Trichechus manatus*). Abstr. 10th Bien. Conf. Biol. Mar. Mamm., Galveston, TX, Nov. 1993:53, 130 pp.

Hall, J.D., and C.S. Johnson. 1971. Auditory thresholds of a killer whale Orcinus orca Linnaeus. J. Acoust. Soc. Am. 51:515–517.

Hartman, D.S. 1979. Ecology and behavior of the manatee (*Trichechus manatus*). *In*: Florida Spec. Publ. Am. Soc. Mammal. 5, 153 pp.

Johnson, C.S. 1968. Relation between absolute hearing threshold and duration-of-tone pulses in the bottlenosed porpoise. J. Acoust. Soc. Am. 43(4):757–763.

Johnson, S.R., J.J. Burns, C.I. Malme, and R.A. Davis. 1989. Synthesis of information on the effects of noise and disturbance on major haulout concentrations of Bering Sea pinnipeds. OCS Study MMS 88–0092. Rep. from LGL Alaska Res. Assoc. Inc., Anchorage, AK, for U.S. Minerals Manage Serv., Anchorage, AK. 267 pp. NTIS PB89-191373.

Kastelein, R.A., C.L. van Ligtenberg, I. Gjertz, and W.C. Verboom. In press. Free field hearing tests on wild Atlantic Walruses (*Odobenus rosmarus*) in air. Aquat. Mammals 19(3).

Ketten, D.R. 1991. The marine mammal ear: specializations for aquatic audition and echolocation. *In*: D.B. Webster, R.R. Fay, and A.N. Popper (eds.), Evolutionary Biology of Hearing. Springer-Verlag, Berlin. pp. 717–750.

Ketten, D.R., D.K. Odell, and D.P. Domning. 1992. Structure, function, and adaptation of the manatee ear. *In*: J.A. Thomas, R.A. Kastelein, and A.Y. Supin (eds.), Marine Mammal Sensory Systems. Plenum, New York. pp. 77–95.

Leggat, L.J., H.M. Merklinger, and J.L. Kennedy. 1981. LNG carrier underwater noise study for Baffin Bay. *In*: N.M. Peterson (ed.), The Question of Sound from Ice-breaker Operations: The Proceedings of a Workshop. Arctic Pilot Proj., Petro-Canada, Calgary, Alb. pp. 115–155.

LGL and Greeneridge. 1987. Responses of bowhead whales to an offshore drilling operation in the Alaskan Beaufort Sea, autumn 1986. Rep. from LGL Ltd., King City Ont., and Greeneridge Sciences, Inc., Santa Barbara, CA, for Shell Western E & P. Inc., Anchorage, AK. 371 pp.

Ljungblad, D.K., B. Würsig, S.L. Swartz, and J.M. Keene. 1988. Observations on the behavioral responses of bowhead whales (*Balaena mysticetus*) to active geophysical vessels in the Alaskan Beaufort Sea. Arctic 41(3):183–194.

Malme, C.I., P.R. Miles, C.W. Clark, P. Tyack, and J.E. Bird. 1983. Investigations of the potential effects of underwater noise from petroleum industry activities on migrating gray whale behavior. BBN Rep. 5366. Rep. from Bolt Beranek & Newman Inc., Cambridge, MA, for U.S. Minerals Manage. Serv., Anchorage, AK. Var. pag. NTIS PB86-174174.

Malme, C.I., P.R. Miles, C.W. Clark, P. Tyack, and J.E. Bird. 1984. Investigations of the potential effects of underwater noise from petroleum industry activities on migrating gray whale behavior/Phase II: January 1984 migration. BBN Rep. 5586. Rep. from Bolt Beranek & Newman Inc., Cambridge, MA, for U.S. Minerals Manage. Serv., Anchorage, AK. NTIS PB86-218377.

Malme, C.I., B. Würsig, J.E. Bird, and P. Tyack. 1988. Observations of feeding gray whale responses to controlled industrial noise exposure. *In*: W.M. Sackinger, M.O. Jeffries, J.L. Imm, and S.D. Treacy (eds.), Port and Ocean Engineering Under Arctic Conditions, vol. II. Geophysical Inst., Univ. of Alaska, Fairbanks, AK. pp. 55–73.

Miles, P.R., C.I. Malme, and W.J. Richardson. 1987. Prediction of drilling site-specific interaction of industrial acoustic stimuli and endangered whales in the Alaskan Beaufort Sea. BBN Rep. 6509; OCS Study MMS 87–0084. Rep. from BBN Laboratories Inc., Cambridge, MA, and LGL Ltd., King City, Ont., for U.S. Minerals Manage. Serv., Anchorage, AK. 341 pp. NTIS PB88-158498.

Møhl, B. 1968. Auditory sensitivity of the common seal in air and water. J. Aud. Res. 8:27–38.

Moore, P.W.B., and R.J. Schusterman. 1987. Audiometric assessment of northern fur seals, *Callorhinus ursinus*. Mar. Mamm. Sci. 3(1):31–53.

Mulroy, M.J. 1991. Munk's experiment. Science 253:118–119.

Myrberg, A.A., Jr. 1978. Ocean noise and the behavior of marine animals: Relationships and implications. *In*: J.L. Fletcher and R.-G. Busnel (eds.), Effects of Noise on Wildlife. Academic Press, New York. pp. 169–208.

Nishiwaki, M., and H. Marsh. 1985. Dugong. *In*: S.H. Ridgway and R.J. Harrison (eds.), Handbook of Marine Mammals. Vol. III, The Sirenians and Baleen Whales. Academic Press , London. pp. 1–31.

Payne, R., and D. Webb. 1971. Orientation by means of long range acoustic signaling in baleen whales. Ann. N.Y. Acad. Sci. 188:110–141.

Popov, V., and A. Supin. 1990. Electrophysiological studies of hearing in some cetaceans and a manatee. *In*: J.A. Thomas and R.A. Kastelein (eds.), Sensory Abilities of Cetaceans/ Laboratory and Field Evidence. Plenum, New York. pp. 405–415.

Reeves, R.R. 1977. The problem of gray whale (*Eschrichtius robustus*) harassment: At the breeding lagoons and during migration. U.S. Mar. Mamm. Comm. Rep. MMC-76/06. NTIS PB-272506. 60 pp.

Reeves, R.R. 1992. Whale responses to anthropogenic sounds: a literature review. Sci. & Res. Ser. 47. N.Z. Dep. Conserv., Wellington, New Zealand. 47 pp.

Richardson, W.J., and C.I. Malme. 1993. Man-made noise and behavioral responses. *In*: J.J. Burns, J.J. Montague, and C.J. Cowles (eds.), The Bowhead Whale. Spec. Publ. 2, Soc. Mar. Mamm., Lawrence, KS. pp. 631–700.

Richardson, W.J., M.A. Fraker, B. Würsig, and R.S. Wells. 1985. Behaviour of bowhead whales *Balaena mysticetus* summering in the Beaufort Sea: reactions to industrial activities. Biol. Conserv. 32(3):195–230.

Richardson, W.J., B. Würsig, and C.R. Greene, Jr. 1986. Reactions of bowhead whales, *Balaena mysticetus*, to seismic exploration in the Canadian Beaufort Sea. J. Acoust. Soc. Am. 79 (4):1117–1128.

Richardson, W.J., B. Würsig, and C.R. Greene, Jr. 1990. Reactions of bowhead whales, *Balaena mysticetus*, to drilling and dredging noise in the Canadian Beaufort Sea. Mar. Environ. Res. 29(2):135–160.

Richardson, W.J., C.R. Greene, Jr., C.I. Malme, and D.H. Thomson. 1991. Effects of noise on marine mammals. OCS Study MMS 90–0093; LGL Rep. TA834-1. Rep. from LGL Ecol. Res. Assoc. Inc., Bryan, TX, for U.S. Minerals Manage. Serv., Atlantic OCS Reg., Hemdon, VA. NTIS PB91-168914. 462 pp.

Ridgway, S.H., and D.A. Carder. 1993. High-frequency hearing loss in old (25+ years old) male dolphins. J. Acoust. Soc. Am. 94(3, Part 2):1830.

Riedman, M.L. 1984. Effects of sounds associated with petroleum industry activities on the behavior of sea otters in California. Appendix D . *In*: Malme et al. (1984). NTIS PB86-218377. Var. pag.

Ross, D. 1976. Mechanics of Underwater Noise. Pergamon, New York. 375 pp.

Schusterman, R.J., R.F. Balliet, and J. Nixon. 1972. Underwater audiogram of the California sea lion by the conditioned vocalization technique. J. Exp. Anal. Behav. 17(3):339–350.

Simmonds, M.P., and L.F. Lopez-Jurado. 1991. Whales and the military. Nature 351:448.

Spindel, R.C., and P.F. Worcester. 1990. Ocean acoustic tomography. Sci. Am. 263(4):94–99.

Tavolga, W.N. 1967. Noisy chorus of the sea. Natural History 76:20–27.

Terhune, J.M., and K. Ronald. 1972. The harp seal, Pagophilus groenlandicus (Erxleben, 1777). III. The underwater audiogram. Can. J. Zool. 50:565–569.

Terhune, J.M., and K. Ronald. 1975. Underwater hearing sensitivity of two ringed seals (*Pusa hispida*. Can. J. Zool. 53:227–231.

Thomas, J., N. Chun, W. Au, and K. Pugh. 1988. Underwater audiogram of a false killer whale (*Pseudorca crassidens*). J. Acoust. Soc. Am. 84(3):936–940.

Urick, R.J. 1982. Sound Propagation in the Sea. Peninsula Publishing, Los Altos, CA. Var. pag.

Urick, R.J. 1983. Principles of Underwater Sound, 3rd ed. McGraw-Hill, New York. 423 pp.

Urick, R.J. 1986. Ambient Noise in the Sea. Peninsula Publishing, Los Altos, CA. Var. pag.

Weigle, B.L., I.E. Beeler-Wright and J.A. Huff. 1993. Responses of manatees to an approaching boat: a pilot study. Abstr. 10th Bien. Conf. Biol. Mar. Mamm., Galveston, TX, Nov. 1993:111. 130 pp.

Wenz, G.M. 1962. Acoustic ambient noise in the ocean: Spectra and sources. J. Acoust. Soc. Am. 34 (12):1936–1956.

White, M.J., Jr., J. Norris, D. Ljungblad, K. Baron, and G. di Sciara. 1978. Auditory thresholds of two Beluga whales (*Delphinapterus leucas*). Hubbs/Sea World Res. Inst. Tech. Rep. 78–109 for Naval Ocean Systems Center , San Diego, CA. 35 pp.

Worcester, P.F., J.F. Lynch, W.M.L. Morawitz, R. Pawlowicz, P.J. Sutton, B.D. Cornuelle, O.M. Johannessen, W.H. Munk, W.B. Owens, R. Shuchman, and R.C. Spindel. 1993. Evolution of the large-scale temperature field in the Greenland Sea during 1988–1989 from tomographic measurements. Geophys. Res. Lett. 20(20):2211–2214 (22 Oct. 1993).

2

Regulatory Barriers and Possible Alternatives

Chapter 1 focused on the need for additional information so that the effects of low-frequency sound on marine mammals might be predicted and the potential trade-offs between the benefits of research that uses low-frequency sound and the costs to marine mammals might be assessed. As Chapter 3 indicates, the research that would provide some of the missing information is conceptually straightforward biological science, the proposed experiments should provide much of the needed information, and the cost is not enormous compared with that of other scientific efforts of comparable magnitude. However, research on marine mammals is heavily regulated and often delayed—even more than research on human subjects—which makes the prospects for timely conduct of some of these much-needed studies poor. This chapter discusses these regulatory burdens and suggests changes that would enable legitimate and needed research to proceed.

THE MARINE MAMMAL PROTECTION ACT AND ITS INTERPRETATION

The Marine Mammal Protection Act (MMPA) was enacted by Congress in 1972, because species and populations of marine mammals were in "danger of extinction or depletion as a result of man's activities" (P.L. 92–522, Sect. 2(1)—the first finding stated by Congress in the MMPA as originally passed in 1972). The central component of the act was "a moratorium on the taking or importation of marine mammals" (Sect. 101). Certain exemptions to the moratorium were permitted: the two most important with regard to the potential impact of low-frequency sound are (1) permits for scientific research that would directly study or benefit marine mammals, and (2) "small incidental take" (SIT) authorizations for human activities that may disturb or even kill small numbers of marine mammals but that have negligible effects on the overall population. Few SIT authorizations have been applied for because of the complexity and time-consuming nature of the authorization process. Those that have been granted for noise-producing activities were for offshore drilling, seismic exploration, and rocket launches in a few specific coastal regions around the United States. In addition to activities such as those, other types of sound-related scientific studies that do not directly benefit marine mammals are hydroacoustic studies of plankton distribution or acoustic tomography studies of ocean currents. Such studies have proceeded without SIT authorization in the past, but are now coming under increasing regulatory scrutiny and scrutiny by the public and environmental groups. Again, we point out that what are called scientific research permits are only issued for research *on or directly benefiting* marine mammals; general scientific experiments that disturb or harm marine mammals are required to apply for SIT authorizations. The issues involved with each of these types of permit are discussed in the following sections of this chapter.

The U.S. National Marine Fisheries Service (NMFS) is responsible for enforcing the moratorium on the "taking" of most types of marine mammals and for granting exemptions. The U.S. Fish and Wildlife Service (FWS) is responsible for walruses, polar bears, sea otters, and manatees. The definition of "take" in the MMPA (Definition #12) is "to harass, hunt, capture or kill, or attempt to harass, hunt, capture or kill any marine mammal." What constitutes harassment is not defined in the MMPA.[1]

Of all human activity, commercial fishing kills the largest numbers of marine mammals [see National Research Council (1992) for information on dolphin mortality associated with tuna fishing]. Since 1988, commercial fisheries within the jurisdiction of the United States have operated under a blanket exemption from the MMPA and can kill marine mammals, even from depleted populations, as long as

[1] This chapter was written before the 1994 amendments to the MMPA were finalized.

they register and report all animals killed.[2] Direct hunting by native peoples is also allowed. Large numbers of fur seals are killed in legal hunts in the north Pacific, and various marine mammals are killed in legal subsistence harvests in Alaska.

Although as an activity scientific research has comparatively little impact on marine mammal populations, it is the object of considerable regulation. For example, according to an annual NMFS report on MMPA, "One of the most extensive administrative programs in NMFS is the permit system that authorizes the taking of marine mammals for scientific research" (National Marine Fisheries Service, 1985). For reasons discussed in this chapter, it is the committee's judgment that this regulatory system actively discourages and delays the acquisition of knowledge that would benefit conservation of marine mammals, their food sources, and their ecosystems (see discussion in Le Boeuf and Würsig, 1985, p. 143). If strictly enforced, current procedures would also actively discourage some types of biological research (described below) that could be of great benefit to humans. Current procedures also impede the research of physical scientists who use high-intensity sound sources. Such studies provide knowledge about underwater acoustics, physical oceanography, and geophysics.

HARASSMENT OF MARINE MAMMALS

One of NMFS's mandated responsibilities is to deal with actions that might take marine mammals by harassment—including actions involving the introduction of sounds into the water, because the definition of "taking" in the MMPA includes harassing. However, the definition of harassing is included neither in the MMPA nor in the regulations. [3] Logically, the term harassment would refer to a human action that causes an adverse effect on the well-being of an individual animal or (potentially) a population of animals. However, "the term 'harass' has been interpreted through practice to include any action that results in an observable change in the behavior of a marine mammal...." (Swartz and Hofman, 1991). In the opinion of the committee, this interpretation is inappropriate.

There is no scientific evidence to indicate that the relatively mi-

[2] This commercial fishing interim exemption from the moratorium on "taking" a marine mammal under the Marine Mammal Protection Act expires April 1, 1994.

[3] NMFS has proposed a definition of harassment in the proposed regulations published in the Federal Register last year.

nor and short-term behavioral reactions observed in marine mammals in many field situations have any long-term, adverse effect on the animals. Marine mammals, like other animals, show alerting responses and orient to many stimuli, natural and human-made. These reactions are part of their normal behavioral repertoire, and may or may not indicate an adverse effect on the animal.

As discussed in Chapter 1, for example, in gray and bowhead whales, subtle behavioral changes have been detected through detailed analysis of migration paths. Sounds barely audible in the background noise of the sea can cause migrating whales to deflect a few hundred meters around the stationary sound source (Malme et al., 1983, 1984). One could argue that such brief and energetically trivial behavioral changes do not harm the whales and, in fact, prevent them from physically encountering a potential disturbance. Of course, stronger sources may ensonify a larger area, in which case the changes in migrating paths are more extensive and the potential disturbance may be more extreme (LGL & Greeneridge, 1987; Brewer et al., 1993).

As researchers develop more sophisticated methods for measuring the behavior and physiology of marine mammals in the field (e.g., via telemetry), it is likely that detectable reactions, however minor and brief, will be documented at lower and lower received levels of human-made sound. It is conceivable that, with sufficient effort, subtle reactions might be detectable at any sound level intense enough to be heard by the animal (absolute detection threshold). These reactions may not involve a disruption of normal behavior or displacement. In that case, subtle and brief reactions are likely to have no effect on the well-being of marine mammal individuals or populations. Just as it would not be sensible to extend the definition of harassment to each successively lower level at which some subtle new reaction is detected, it is not sensible to have it set now at a level that activates alerting responses or minor course deviations. It is difficult to believe that these minor and short-term responses have any profound long-term influence on mammals. Indeed, such responses are probably beneficial to the species by allowing the animals to keep a safe distance from potential harm associated with some sound sources such as drilling equipment and moving ships.

In humans, we draw clear distinctions among sounds that are detectable, annoying, and physically harmful. Levels that are annoying or harmful are highly variable, but are generally considered to be far above those that are barely detectable. OSHA does not require that special measures be taken to protect the hearing of workers unless the average sound level in the workplace is above 90 dBA (re 20 μPa—air standard). This level is about 100 dB above the level at which sound might first be detectable for humans. It is reasonable to suppose that a gap also exists between detection and damage thresholds for marine mammals, but current knowledge is inadequate to estimate the size of this difference. Additional research on absolute auditory sensitivity and temporary threshold shift (see Chapter 3) is the only way to acquire this important information. Annoyance thresholds for humans are somewhere between the detection threshold and damage limits, but there is no national standard in the United States. The annoyance thresholds for marine mammals are largely unknown, and it is probable that it will be at least as difficult to establish such standards for marine mammals as it has been for humans.

PERMITTING IMPLICATIONS FOR RESEARCH ACTIVITY

Scientific Research Permits

Scientists who propose to conduct research directed toward marine mammals are aware of the permitting requirements of the MMPA and of the Endangered Species Act (ESA) and the associated regulations. Most of their research can be conducted under the scientific permitting process. They routinely apply for and obtain such scientific research permits. However, the lengthy and unpredictable duration of this process can create serious difficulties for research. For example, most permits are issued within about four months, but there is no guarantee that a response must be made to an applicant within any specific time. It is sometimes difficult to obtain such permits in time to conduct projects that must be completed within fixed time periods. It is often impossible to take advantage of some unforeseeable "window of opportunity" such as a whale stranding. A major drawback in current procedures is that regardless of the application date, permits are sometimes granted only a few days before a study is to begin. In some cases, the delay can be due to a researcher submitting insufficient or incomplete information in the permit application, resulting in the need for the applicant to supply additional information to NMFS before a permit can be issued. In addition to permit delays, certain types of research that are considered "invasive" or "controversial" either are not allowed under the current permitting process or may require an Environmental Assessment or even an Environmental Impact Statement under the National Environmental Protection Act (NEPA). Such a regulatory burden actively discourages researchers from pursuing those lines of study.

The laws governing the "take" of marine mammals, including MMPA, ESA, and NEPA, set many of the requirements that must be met before scientific research permits or incidental take authorizations can be issued. The National Marine Fisheries Service and the Fish and Wildlife Service must implement permit procedures that comply with these laws. The committee recognizes that some of the changes in permitting procedures recommended later in this chapter would require changes in the applicable legislation. However, other changes could be made without legslative action through changes in the regulations that are implemented by NMFS and FWS.

The committee strongly agrees with the objective of marine mammal conservation, but it believes that the present emphasis on regulation of research is unnecessarily restrictive. Not only is research hampered, but the process of training and employing scientists with suitable skills is impeded when research projects cannot go forward. Experienced researchers are the ultimate source for expanding our knowledge of marine mammals. A policy that interferes with the development of this resource appears to be self-defeating.

Not only can current regulatory processes impede the carrying out of specific projects and the training and employing of scientists, but there are perhaps other losses as well, from studies not permitted if they do not hold potential for benefiting marine mammals. For example, marine mammals might be useful in investigations of human or terrestrial mammal medical problems. Such kinds of research might include (1) the study of the control of bradycardia and vasoconstriction (diving response) for application in the sudden infant death syndrome or various cardiac arrhythmias, (2) the study of cetacean bone with possible application for osteoporosis, (3) the study of marine mammal immunology for application to immune system understanding, and (4) the study of the dolphin brain and/or sound-processing system for the understanding of communication disorders.

Furthermore, there are nonmedical studies that might not be seen to have direct benefit to the species, yet still yield important information for humans. Such studies include (5) the investigation of dolphin swimming or hydrodynamics with application in ship hull design, (6) the examination of dolphin echolocation with application in sonar design, and (7) the examination of dolphin neural processing with application in neural networks or computer logic.

Under current procedures, studies such as the seven listed and others similar to them are unlikely to be permitted unless they are non-invasive or promise potential for benefiting marine mammals. Any intrinsic interest or potential application for human or terrestrial mammals regrettably seems not to be embraced by the current permitting process. The original MMPA was not as restrictive in this respect as the law is now. Amendments added in 1988, however, allowed for much more stringent treatment of research by federal regulators. Section 104(b)(3) of the act states: "A permit may be issued for scientific research purposes only to an applicant who submits with its permit application information indicating that the taking is required to further a bona fide scientific purpose and does not involve unnecessary duplication of research." Federal permit officials may take a very restrictive stance on what "bona fide scientific purpose" or "unnecessary duplication" is. Restrictive interpretations of the former phrase can potentially affect or defeat worthwhile research that is not directly on or of benefit to marine mammals, such as the seven types of studies listed above, even if they are nonetheless beneficial and probably do not harm marine mammals in the process. Restrictive interpretations of "unnecessary duplication" can affect research on marine mammals and on other scientific subjects such as those listed. Most scientists feel that detailed peer review, a standard part of good research, should be used to determine what is "bona fide" or duplicative rather than value judgments of permit officials. One suggestion is to amend Section 104(b)(3) of the MMPA (see Heck and Buck, 1992).

Small Incidental Take (SIT) Authorizations

Most biological research in the oceans is not on marine mammals but on other organisms such as fishes and invertebrates. Not only are many of these organisms prey of marine mammals and therefore vital for their survival, but they are essential for a healthy, functioning marine ecosystem. Moreover, much of the research on these organisms is used for understanding and thereby regulating sustainable fisheries. Some of the biological research on these organisms, for example, studies of krill populations in the deep scattering layers, directly employs acoustical tools (Mauchline, 1980). Also, the research vessels generate noise and, by being in marine mammal habitats, their activities can disrupt marine mammal behavior (e.g., by attracting or frightening the animals).

Even though it might interfere with or affect marine mammals, scientific research on animals other than marine mammals has not been interpreted as being eligible for scientific research permits under the Marine Mammal Protection Act. These permits, in comparison with the SIT authorizations, are less difficult and time-consuming. However, as stated above, the MMPA is interpreted to require that eligible research be on or benefiting marine mammals. Thus, a researcher wishing to ensure that a study of animals other than marine mammals is in compliance with U.S. law would need to apply for a small incidental take authorization. However, this is such a complex and time-consuming process that the application is considered to be a major undertaking, even by the few industrial groups and governmental agencies that have gone ahead with this process. The process may require as much as two years to complete, and may even require public hearings. As presently implemented, the SIT process is impractical for the typical individual researcher or for most small research teams. As part of that application, for example, the researcher must specify the geographic area where the work is to be done and the species and numbers of marine mammals that will be "taken." Trying to anticipate how such research might modify the behavior of a marine mammal at various distances, however slight the modification, is often an impossible task. Furthermore, few researchers or funding agencies can tolerate the delay of at least one and possibly two years between the formulation and the execution of a research plan. And clearly, the extra cost of hiring marine mammalogists to prepare the application and to monitor the effects of this research would present another obstacle.

Likewise, in physical oceanography and marine geology and geophysics, much of the research employs powerful low-frequency acoustic sources that have a broad range of frequencies and intensities. Consequently, there is a strong probability that such research will modify the behavior of some marine mammals and hence constitute a "take" under the present interpretation of the MMPA. Loss of such research would have a profound effect on several areas of the physical sciences, and would deny access to important information about Earth. Geophysics uses acoustic information provided by low-frequency sources to determine the profile of the ocean floor and to make inferences about plate tectonics. Shifts in these plates result in earthquakes. Ocean acoustic tomography measures acoustic transmissions over very long paths—of the order of Earth's diameter. The travel times along these paths can be measured precisely, and from such measurements the velocity of propagation (which is temperature-dependent) can be calculated. Such experiments have the potential for estimating the average temperature in the ocean over great areas. This work is actively being pursued to provide information about global warming. The field of ocean acoustics is concerned with understanding propagation in ocean waveguides and reflections from the surface and bottom of the sea. Information from this area has played an important role in the development of sonar and other methods of underwater sensing. Basic research of this kind could also be helpful in improving acoustic systems that are capable of tracking marine mammals and their food supply.

Not only is the SIT authorization process for basic research complicated and time-consuming, but the *small* incidental take provisions of the MMPA allow only for cases in which a small number of marine mammals will be subject to a negligible effect. It does not allow for cases in which a larger number might be affected in a negligible manner. The latter cases could arise when projected sounds were detectable over a large area at relatively low levels. At present, it is doubtful whether some acoustic oceanography projects could legitimately obtain SIT authorizations. They might not satisfy the small-number criterion, and the requirement to monitor the marine mammals potentially subject to a "take" would be difficult, if not impossible, to satisfy in any meaningful manner.

In the past, scientific studies on topics other than marine mammals have proceeded without SIT authorizations, even if there was a strong probability that marine mammals would be affected. Such studies are coming under increasing regulatory scrutiny. If the present SIT process is applied to the wider spectrum of scientific projects, the complexities and delays of the process, along with the limitation to small takes, would have a chilling effect on a wide area of scientific enterprise.

The committee believes that broadening the MMPA definition of scientific research permits to include areas of science not directly related to marine mammals would greatly facilitate many studies of ocean science. If the SIT process must be used for some or all of these studies, then the reinterpretation of "small" to include studies in which many animals might be affected, but by a negligible amount, would greatly aid many legitimate scientific research studies conducted in the ocean.

MAGNITUDE OF POTENTIAL NOISE POLLUTION FROM OCEANIC ACOUSTIC STUDIES

Although the amount of noise pollution created by the scientific enterprise varies greatly depending on the nature of the research, it should be kept in mind that it is only one of the many seafaring activities of the human species. Appendix C presents the committee's calculations of how the amount of noise pollution generated by typi-cal oceanographic research studies compares with that produced by another common source of ocean noise, namely, the supertanker. If we restrict our concerns to a frequency band near 50 Hz, a band often used in oceanographic experiments, the yearly acoustic energy generated by the fleet of supertankers is about *50* times that generated by a year's worth of oceanographic experiments. If a wider frequency band was included the ratio would be more than 1,000 to 1, because the supertanker generates noise over a wider frequency band than the oceanographic experiment does.

The analysis in Appendix C is based entirely on a calculation of the overall energy of the two types of sources. It does not take into account the location of the source and the marine mammals or the transmission path between the two. It also does not include the frequency composition of the source, which could be of great importance for some species. Also, if the oceanographic experimenter placed a source in the SOFAR (Sound Fixing and Ranging) channel (the layer of seawater from about 700 meters down to 1,500 meters in which sound travels relatively slowly) it would likely ensonify a much greater area than would a supertanker, whose sounds are generally poorly coupled to the SOFAR channel. In higher latitudes, however, the supertanker generally couples well with the SOFAR channel, which is close to the surface at those latitudes.

In evaluating the potential effect of a specific acoustic exposure on marine mammals, one would need to consider the specific propagation conditions, the sound spectra, the marine mammals, and members of the food chain present. Given the present lack of knowledge, the advisable course is, when feasible, to locate and schedule all scientific exposures so as to avoid areas and seasons in which high densities of marine mammals might be present. In particular, the experiments should, when feasible, avoid feeding, breeding, or rearing areas at seasons when these areas are critical to the animals.

This discussion of potential noise pollution should not be interpreted as approving or encouraging sonic pollution of a scientific origin or as justifying it in terms of another source of pollution. Indeed, on an evolutionary time scale, supertankers are quite a recent phenomenon. Some might argue that supertankers could be producing major negative effects on marine mammals and their food chain, and that evidence of this is yet to emerge. Whatever the case, ocean noise pollution will not be greatly reduced by meticulously regulating ocean acoustic studies, a class of sources that contributes less than 2 percent of the current yearly energy near 50 Hz. What is badly needed is more scientific research to assess the effects produced by all noise sources. This research would provide the knowledge that is needed for understanding the potential impact of all such pollutants.

PROPOSED CHANGES IN THE REGULATORY STRUCTURE

The committee considered several possible alternatives for facilitating valuable research while maintaining all necessary protection for marine mammals. The committee notes that Section 101(a)(3) of the Marine Mammal Protection Act permits waivers of the act's provisions by the Secretary of Commerce and the Secretary of the Interior; we see no acceptable means for using the waivers to solve the current regulatory difficulties. The following alternatives seem worthy of consideration.

Using a New Mechanism to Regulate Scientific "Takes"

NMFS has proposed a new regime to focus on the larger purpose of maintaining all marine mammal populations at optimum sustainable population levels (National Marine Fisheries Service, 1992). If adopted by Congress, this regime would calculate for each species a conservative number of animals that might be "taken" by all human activities, and then allocate this potential biological removal (PBR) to different human user groups.

The proposed regime is designed to redirect regulation to focus on human activities with the largest impact on marine mammal populations, scaling the extent of regulation to the risk the activity poses to populations. The proposed regime was initially developed primarily for commercial fishing, but it was designed to allow the inclusion of other "user groups" for PBR. If such a mechanism is adopted in the revised legislation, this committee recommends that Congress and NMFS consider it for regulating most "takes" of marine mammals by research as well. Since the objective of the law is to protect marine mammals, it is difficult to understand applying different, and less stringent, rules to activities that kill marine mammals than to activities that are known to benefit them or to have negligible effects on them. For any population in which harassment is considered to be a serious risk to populations, taking by harassment may be included in these regulations. Where taking by fishing is considered to have negligible impact compared to other activities, regulatory attention should focus on these more significant risks.

Broadening the user groups in these regulations would improve NMFS's ability to monitor the effects of human activities in a more comprehensive manner. All takes would be monitored through a permit system requiring reporting of takes. This alternative permit system would be more practical than that currently applied to marine mammal research, and it could also be applied to other kinds of scientific research.

Utilization of the IACUC System

Another alternative scheme of regulation would be to make use of institutional animal care and use committees (IACUCs). In the United States, scientific research involving nonendangered terrestrial, and sometimes marine, mammals is reviewed locally by these committees. They operate under federal agency rules and guidelines and approve or disapprove protocols for research projects submitted by scientists for all animal-related research within the institutional purview.

An IACUC provides peer review, and each committee also has an outside member representing the general public. The IACUCs have standard procedures for dealing with research on endangered species. It should be noted, however, that IACUCs cannot issue endangered species or marine mammal research permits under existing law. The IACUC system avoids the time delays of any process centered in Washington, D.C.

IACUCs are effective in protecting individual animals. The focus on the individual makes it uncertain whether this approach would be sufficient to protect populations. At a minimum, NMFS or FWS or both would have to work with the IACUCs to provide them with timely information about population levels and geographic areas of importance to the animals for use by the IACUCs during proposal review and approval.

Streamlining Existing Regulations

If a system of regulation similar to the existing one must be maintained, then this committee suggests that the scientific research permit process be revised or streamlined. Some of these suggestions may require changes in the existing legislation. The following steps are appropriate whether or not one or more of the alternative regulatory approaches described above is implemented for some research situations. The suggested actions are these:

1. Distinguish between different types of "taking," for example, incidental disturbance effects versus "hunt, capture, collect or kill." Streamline the permitting process for projects in which marine mammals would not be killed or captured. Further, streamline the process for projects in which the effects would clearly be not only nonlethal but also negligible. Streamlining could include some or all of the following:

a. Require less information and/or fewer details in permit application.
b. Make provision for noncontroversial applications to be approved without time-consuming review outside the specific permitting authority.
c. Allow permits to cover a broader range of circumstances (e.g., more species, wider area, wider range of study protocols) without requiring a permit modification or new permit.

2. Establish a set schedule for the processing of applications for scientific research permits. Obtain the staff needed to allow the schedule to be met. For example, within a specified period of time after receipt of a permit application (e.g., 10 days), the responsible agency (NMFS or FWS) should complete the "initial review" phase,[4] that is, determine what action is appropriate and take that action: return to applicant for more information, determine that an environmental assessment (EA) is required under the National Environmental Policy Act of 1969 and initiate that process, or publish the required notice in the *Federal Register* and pass the application to the Marine Mammal Commission for comment. The public comment period (presently 30 days) should be nonextendable. Within a set time after expiration of that comment period (e.g., 10 days), the agency should issue or deny the permit, or advise the applicant that there are sufficient questions to require some other action such as a public hearing or EA. This approach would provide applicants with some assurance that a well-prepared and noncontroversial application would be processed within a set period of time.
3. Amend the MMPA explicitly to allow scientific research permits to be issued for scientific research that is not specifically directed toward the study or benefit of marine mammals, provided the research meets reasonable standards. These could include standards or components similar to those that marine mammal studies must meet to obtain a research permit and that other human activities must meet to qualify for an incidental take authorization. For example, the research might be required to have some or all of the following attributes: to be nonlethal, to have negligible anticipated effects on individual marine mammals and their populations, to include an approved marine monitoring program where appropriate, to pass an acceptable process of peer review within its own discipline, and to be "not unnecessarily duplicative."

[4] The revised permit regulation recently proposed by NMFS does not set any time limit for the 'initial review' phase (Federal Register, 14 October 1993, p. 53341).

The committee also suggests that the incidental take authorization process be streamlined to make it practical for scientific researchers to use where appropriate. Insofar as possible, while maintaining adequate protection for marine mammals, this streamlining should include a reduced level of detail in the application for incidental take authorizations, more rapid processing, and clarification of the nature and extent of the marine mammal monitoring (if any) that would be required. Once authorizations are approved, applications for project-specific letters of authorization should require minimal lead time, if they are necessary at all.

The committee further suggests that consideration be given to a revision in the incidental take provisions of the MMPA to delete references to *small* numbers of marine mammals. Such a change would allow NMFS or FWS to approve an incidental take authorization for any situation in which predicted effects on individuals and populations were negligible, regardless of the numbers of marine mammals that might be involved. This procedure could, for example, make an incidental take authorization possible in the case of an acoustical oceanography project that would ensonify an area large enough that more than a "small" number of marine mammals might be exposed to significant noise, provided that it could be demonstrated that the effect of the sound on these animals would be negligible. The committee recognizes that changes in the incidental take authorization process might also broaden its application to other types of human activity. However, it seems unreasonable that an exemption from the "take" prohibitions of the MMPA should be available for some human activities, including some that kill many marine mammals, without being available for other human activities whose goal may include the acquisition of information of potential value for the conservation of marine mammals.

SUMMARY OF RECOMMENDED CHANGES

The committee recommends altering the regulatory structure to facilitate legitimate research on low-frequency sound and its effects on marine mammals in the following ways:

- Broaden the definition of research for which scientific permits can be issued to include research activities beyond those "on or directly benefiting marine mammals." The population status of the species and the kind of "take" should determine the number of allowable takes, and the same regulations should apply equally to all seafaring activities.

- Expedite the regulatory process for activities involving nonlethal takes and further simplify the process for those nonlethal takes producing only negligible impact.
- Reword the incidental take authorization to delete references to "small" numbers of marine mammals, provided the effects are negligible.
- Consider transferring some aspects of the regulatory process to less centralized authorities patterned after the IACUCs that regulate animal care and safety in the academic and industrial settings.

REFERENCES

Brewer, K. D. M.L. Gallagher, P.R. Regos, P.E. Isert, and J.D. Hall. 1993. ARCO Alaska Inc. Kuvlum #1 exploration prospect/Site specific monitoring program final report. Rep. from Coastal and Offshore Pacific Corp., Walnut Creek, CA, for ARCO Alaska Inc. and the Nat. Mar. Fish. Serv., Anchorage, AK. 80 pp.

Heck, J., and E. Buck. 1992. The Marine Mammal Protection Act: Reauthorization Issues. Congressional Research Service. 92–728 ENR. Library of Congress. Washington, DC. 22 pp.

Le Boeuf, B. J., and B. Würsig. 1985. Beyond bean counting and whale tales. Marine Mamm. Sci. 1 (2):128–148.

LGL and Greeneridge. 1987. Responses of bowhead whales to an offshore drilling operation in the Alaskan Beaufort Sea, autumn 1986. Rep. from LGL Ltd., King City, Ont., and Greeneridge Sciences Inc., Santa Barbara, CA, for Shell Western E & P. Inc., Anchorage, AK. 371 pp.

Malme, C.I., P.R. Miles, C.W. Clark, P. Tyack, and J.E. Bird. 1983. Investigations of the potential effects of underwater noise from petroleum industry activities on migrating gray whale behavior. BBN Rep. 5366. Rep. from Bolt Beranek & Newman Inc., Cambridge, MA, for U.S. Minerals Manage. Serv., Anchorage, AK. Var. pag. NTIS PB86-174174.

Malme, C.I., P.R. Miles, C.W. Clark, P. Tyack and J.E. Bird. 1984. Investigations of the potential effects of underwater noise from petroleum industry activities on migrating gray whale behavior/Phase II: January 1984 migration. BBN Rep. 5586. Rep. from Bolt Beranek & Newman Inc., Cambridge, MA, for U.S. Minerals Manage. Serv., Anchorage, AK. NTIS PB86-218377.

Mauchline, J. 1980. The Biology of Mysids and Euphausiids. Advances in Marine Biology, vol. 18. 681 pp.

National Marine Fisheries Service. 1985. Marine Mammal Protection Act of 1972: Annual Report 1984/1985. U.S. Department of Commerce, National Oceanic and Atmospheric Administration, Washington, DC. p. 4.

National Marine Fisheries Service. 1992. Proposed Regime to Govern Interactions Between Marine Mammals and Commercial Fishing Operations. U.S. Department of Commerce, National Oceanic and Atmospheric Administration, Washington, DC. 96 pp.

National Research Council. 1992. Dolphins and the tuna Industry. National Academy Press, Washington, DC. 176 pp.

Swartz, S.L., and R.J. Hofman. 1991. Marine Mammal and Habitat Monitoring: Requirements; Principles; Needs; and Approaches. Report prepared for Marine Mammal Commission, August 1991. PB91-215046. pp. 2–3.

3

Topics for Future Research

As discussed earlier in this report, data on the effects of low-frequency sounds on marine mammals are scarce. Although we do have some knowledge about the behavior and reactions of certain marine mammals in response to sound, as well as about the hearing capabilities of a few odontocete and pinniped species, the data are extremely limited and cannot constitute the basis for informed prediction or evaluation of the effects of intense low-frequency sounds on any marine species.

In this chapter the committee identifies several areas in which more research is needed in order to provide a better understanding of the effects of low-frequency sounds on marine mammals and their prey. The next two major sections—Behavior of Marine Mammals in the Wild, and Structure and Function of the Auditory System—describe several of the proposed studies that focus on the acoustic behavior, disturbance responses, and hearing of marine mammals. Marine mammals are not the only marine species that may be affected by intense low-frequency sounds. Although other marine species are outside the direct charge of this committee, many are part of the food chain for marine mammals. If low-frequency sounds change the behavior of, or damage, organisms in the food chain of marine mammals, they may significantly alter the ability of marine mammals to survive, as discussed in the third section of this chapter, Effects of Low-frequency Sound on the Food Chain. The final section, on Development and Applications of Measurement Techniques, describes a number of tools that need to be developed so the basic data that are needed can be obtained.

The intent here is to identify general research needs that are crucial to a full evaluation of the effects of intense low-frequency sounds on a variety of marine mammals and their prey; accordingly, specific experiments are not described in detail.

The order of presentation of topics in the following sections is not meant to suggest a hierarchy of importance. Indeed, any such hierarchy depends upon the goals of the policy makers. For example, if the overriding concerns were with potential damage to the auditory systems of species by low-frequency sound, the most important studies might be those described under the heading Temporary Threshold Shift in the section on the Auditory System. However, if the overriding concern is to identify the risks posed for survival of a specific species by disruption of its ecological niche, research of the sort described in the section Effects of Low-frequency Sound on the Food Chain would have high priority. This committee believes that all of the recommendations described below need to be addressed in order to find out how low-frequency sounds affect marine mammals.

BEHAVIOR OF MARINE MAMMALS IN THE WILD

Aims: To determine the normal behaviors of marine mammals in the wild and their behavioral responses to human-made acoustic signals.

Rationale: There is a substantial lack of knowledge about the normal behaviors of most marine mammals, the role of natural sounds in their lives, and their responses to human-made acoustic stimuli. The primary reason for this lack of knowledge is that marine mammals spend much of their time below the water surface and often at depths that have been, and commonly still are, generally inaccessible to humans. Still, it is crucial to acquire appropriate baseline data on various behavioral dimensions so that any future behavioral changes can be evaluated adequately. The purpose of the proposed studies is to document the normal behaviors of marine mammals and their responses to natural and human-made stimuli.

In the four subsections that follow, the committee suggests several studies employing underwater sound playback and related simulation techniques as methods for testing the reactions of some marine mammals to human-made noise. Observations of reactions to actual human activities are also very desirable. However, playbacks or other simulations are often more practical and have some advantages. For example, (1) It is uncommon for marine mammalogists to have full control over the location and timing of expensive operations such as large ships, seismic vessels, icebreakers, and so forth, especially over a sufficiently long period to obtain meaningful sample sizes. Hence, observations of marine mammals near such operations are usually opportunistic and/or few in number. (2) Although observations near full-scale human activities may be more realistic than those near simulated activities (e.g., playbacks), their very realism can lead to interpretation problems, such as an inability to distinguish acoustic from visual effects. Ideally, controlled or opportune observations near actual human activities would be combined with controlled simulation experiments to take advantage of the complementary advantages and disadvantages of the two approaches (e.g., Richardson et al., 1986).

Natural, Ecologically Important Signals

Aims: To determine how marine mammals utilize natural sounds for communication and for maintaining their normal behavioral repertoire.

It is well known that many marine mammal species utilize sounds for signaling other members of the same species (*intra*specific behaviors) (reviewed in Richardson et al., 1991). For example, mysticetes and some pinnipeds (e.g., Weddell seals) generate and utilize sounds in the normal course of mating and territorial behaviors. Most of these sounds are typically at low to moderate frequencies (tens to thousands of hertz). Odontocetes not only use moderate-frequency sounds for social communication, but they also use higher frequencies (ultrasonic by human standards) for echolocation to detect the presence of objects in their environment. There is also evidence that marine mammals utilize sounds for *inter*specific communication: for example, for detecting predators such as killer whales.

Although there has been extensive research on utilization of sound by captive and coastal animals, less has been done with open-ocean species in the wild. Research on acoustic communication in wild marine mammals has been hindered by problems in following animals and in identifying when a particular animal is producing a sound. Considerable additional work on the ecological significance of natural sounds is essential to any prediction, or evaluation, of stimuli that might disrupt a wide range of behaviors, including those that are sound-dependent. For example, the functions of the low-frequency sounds made by mysticetes and the ranges over which they are used are too poorly understood to infer the biological impact if these signals were masked by noise. Further, once predictable responses to natural acoustic stimuli are identified (e.g., the attraction of animals of the same species by mating calls), this information could be used in designing experiments to determine how other sounds might mask or otherwise change a species's perception of intraspecific signaling.

Investigations on natural sounds of marine mammals might capitalize on existing hydrophone arrays (e.g., the integrated undersea surveillance discussed in the section on Measurement Techniques below), or utilize towed or site-specific arrays. Hydrophone arrays have advantages over single hydrophones: arrays allow localization of the source or sources of calls by triangulation techniques, thereby allowing the matching of received calls with particular animals, their locations, and human activities. Acoustic monitoring should be supplemented, wherever possible, with visual and electronic monitoring so that the behavioral responses to natural or human-made sounds or their interruption can be learned. The ongoing development of increasingly sophisticated tags should be of considerable benefit to this work (see subsection on Tag Development under Measurement Techniques). Accordingly, tag technology deserves increased support by the relevant funding agencies such as the Office of Naval Research, the National Marine Fisheries Service, and the Minerals Management Service.

Habituation to Repeated Human-made Sounds

Aims: To determine the responses of free-ranging marine mammals to human-made acoustic stimuli including repeated exposure of the same individuals. How is the use of natural sounds altered by the presence of human-made sounds?

In addition to the need for research aimed at correlating specific behaviors with passively monitored natural sounds that are produced or used by marine mammals, there is a need for studies in which human-made sounds of various types are presented repeatedly to marine mammals while the responses of the animals are monitored. At present, only very limited data exist on whether sounds having different levels, spectra, or temporal patterns will alter behavior in different ways, and if so, in what species. There are virtually no data on the reactions to repeated exposure to the same sound: that is, do the animals habituate to the sounds? Habituation is the phenomenon of progressive waning of behavioral responsiveness to repeated reception of a stimulus not accompanied by any perceived deleterious effect. Habituation has been observed in many types of animals including marine mammals. Responses of marine mammals to some human-made noises may diminish as the animals learn that these sounds are innocuous. To date, most tests of the reactions of marine mammals to human-made noises have examined only the initial response on first presentation of the noise stimulus. Thus, these results may overestimate the effects of the noises over the long-term (that is, the animal may habituate to the sounds). Clearly, initial responses of marine mammals to sounds may not be indicative of long-term effects. Such studies need to be done over extended periods of time with the same groups of animals.

Where appropriate, tagging should be employed to monitor the behavior of animals before, during, and after ensonification with artificial sounds (including simulations of real-world sounds such as supertankers) and to monitor sound levels adjacent to the animals. Comparison of the responses of known individuals to successive exposures will provide the most important data in these experiments. In effect, each individual will (in part) serve as its own control.

Additional control data of two types should also be obtained, however. (1) Data should be collected in the test area in the year prior to the playback test using the same type of instrumentation that will be used during the playback period (temporal control). (2) Data should also be collected on mammals in a similar control area during both the preplayback and the playback period but without use of the sound stimulation (simultaneous control). Such a design would help eliminate the potentially confounding effects of natural temporal changes and of the presence of observers or tags.

Through the use of telemetry or other tags, investigators will be able to obtain more detailed data on dive history, location, and aspects of animal physiology in response to low-frequency sound signals. The same instrumented individuals should be exposed to the sound on a number of occasions. Such experiments might use a local population of marine mammals engaged in feeding or breeding activity. The committee suggests using sound sources capable of producing sounds of various sorts, including sounds of high intensity and low frequency. A test area of about 10 km square could be ensonified to an intensity level of 126 dB (re 1 μPa—water standard) or greater. Observers on one or more vessels or on shore could observe the animals intermittently. Telemetry receivers on land, in air, or at sea could receive, via radio link, data from tagged animals including ideally the received levels of both the ambient noise and test signals. This type of study might be possible at a location where a suitable acoustic source is to be installed and operated for an acoustical oceanography experiment, thereby making dual use of some logistical arrangements.

The approach described above offers several advantages. (1) Unambiguous measurements could be obtained of the sound-pressure levels received by individual animals. (2) Data could be obtained from several animals over an extended period of time, thereby improving the chances of obtaining statistically reliable effects. (3) Changes in responsiveness over time could be determined (i.e., habituation could be assessed).

Differential Responses of Migrating Whales to Various Human-made Sounds

Aims: To determine how different sound types and levels affect migration and other movement patterns of marine mammals.

Research with gray whales off the coast of central California has demonstrated statistically that received sound levels of 120 dB (re 1 μPa—water standard) can alter paths of migrating whales (Malme et al., 1983, 1984). The paths of these whales are so regular and the numbers of whales so large that it was possible to measure minor course deflections induced by playback of noise stimuli. Such studies suggest that sounds can alter, if only temporarily, the behavior of this species. However, to date, only one impulsive stimulus and five more continuous stimuli have been tested on these animals. This type of experimental method is useful because it provides the opportunity to determine the relative reactions of large samples of baleen whale species to a variety of signals. The work can be done with existing technology, with a high probability of success, and with modest logistical support. Tests of this sort performed with tonal stimuli also have the potential to provide important indirect information about the audiograms in these animals, including data on minimum auditory sensitivity at low frequencies.

Further work with additional stimuli and species would reveal whether the deflections observed in migrating gray whales also occur with sounds having spectral characteristics, durations, and duty cycles other than those already tested. Comparisons should be made between the effects observed with stationary and moving sound sources, and between single and multiple, simultaneous sources.

Responses of Deep-diving Marine Mammals to Low-frequency Sounds

Aims: To determine the responses of deep-diving marine mammals to low-frequency sounds whose characteristics (source level, frequency, bandwidth, duty cycle) duplicate or approximate those produced by acoustic oceanographers.

Sound sources used in acoustic systems designed to monitor ocean warming or for various other ocean acoustical purposes are often placed in the deep sound channel. This channel lies about 700 to 1,500 m below the surface at tropical and temperate latitudes. It is known, or believed, that sperm whales, beaked whales, and elephant seals are capable of diving to depths of 1,000 m or below, and that white whales, pilot whales, bottlenose dolphins, Weddell seals, and other species can dive to depths of at least several hundred meters.

If intense sounds were present during these dives, the animals would encounter large local variations in sound levels that might alter their behavior during diving and feeding. At any given range from the sound source, those species that dive deep enough to enter the sound channel would be exposed to the highest sound levels. Since some of these species may be highly dependent on food obtainable only at great depths, they may be placed at considerable risk by intense low-frequency sounds. At polar latitudes, where the sound channel is often within 200 m of the surface, most species might be influenced by sounds concentrated in the near-surface sound channel. Any changes induced in the behavior of the food species in response to intense sound could also adversely affect marine mammals.

In order to test these possibilities, it would be necessary to deploy one or more sound projectors into the sound channel (when at an accessible depth) in an area where the target species (marine mammals and prey species) occur, and to document the reactions of those animals to the sounds. The projector would probably need to be deployed from a vessel so that it could be moved to locations near tagged animals equipped with data loggers. Behavior should be documented before, during, and after exposure, and the degree of behavioral response should be examined in relation to sound exposure level. It would be desirable to project low-frequency sounds comparable in level and other characteristics to those used by acoustical oceanographers. These sounds should be projected near the depth where operational systems are deployed. This type of study might be done in conjunction with the use of an operational low-frequency acoustic system.

Behavioral data should be obtained for periods while the animals are visible at the surface and while at depth, using instrumentation attached to the animals. From records of position in three dimensions, the positions and tracks of the animals relative to the sound source could be determined before, during, and after exposure to low-frequency sound. Ideally, the recording/telemetry system would also have the ability to determine the sound levels received by each animal under observation.

This type of field study might be most easily implemented on deep-diving pinnipeds such as elephant seals, which are readily tagged when they are out of the water. The ongoing study of sperm whales in deep water close to Dominica (West Indies) supported by the Office of Naval Research already includes many of the elements mentioned above. It appears to have the potential to provide the desired data, but it needs the development of improved recorder/telemetry systems.

STRUCTURE AND FUNCTION OF THE AUDITORY SYSTEM

Aims: To determine the structure and capabilities of the auditory system in marine mammals.

Rationale: To develop an understanding of how human-made signals can affect the behavior of marine mammals, an extensive body of data is needed regarding the basic mechanisms of hearing in these species. Such data are required before informed speculation is possible about whether low-frequency sounds can alter or damage hearing abilities and/or impair the ability of animals to communicate with one another.

Basic Studies of Audiometry

Aim: To determine basic hearing capabilities of various species of marine mammals.

Audiometric functions show the weakest sounds that an animal can hear over the frequency range to which it is sensitive. Although audiometric data do exist for some pinnipeds and several odontocetes (see Figure 2 in Chapter 1), there are very few data on their sensitivity at low frequencies, and there has been little replication within or across species (see subsection on Replication of Data, below in this section). There are no data on hearing abilities of marine mammals as a function of depth—an especially significant question in the case of species that dive to great depths. Also, no data are available on the absolute auditory thresholds of any baleen whale. Still, there is a variety of indirect evidence suggesting that baleen whales are sensitive to low-frequency sounds. For example, large baleen whales produce very low frequency sounds that, in some cases, have been associated with behavioral observations suggesting that the whales can hear their own sounds (Cummings and Thompson, 1971; Clark and Clark, 1980; Thompson et al., 1986).

Two problems with obtaining behavioral data on baleen whales are their size and the attendant difficulty (or impossibility) of working with them in controlled situations. Thus, for the foreseeable future, the only reasonable way to obtain data on these large animals is through use of the evoked potential technique, as described in the next section.

For all species, sensitivity data eventually must be coupled with more detailed studies of masking and critical-band estimates to allow predictions of the minimum signal-to-noise ratios detectable by these animals with particular signals and ambient noise conditions, including low-frequency data. These data, plus knowledge of ocean sound propagation and ambient noise conditions, would allow prediction of the maximum distance at which a given signal might be detectable by an animal.

Measurements on Ensnared or Beached Marine Mammals

Aim: To determine hearing capabilities of larger marine mammals that are not amenable to laboratory study.

As noted above, very little is known about the hearing and auditory systems of large cetaceans such as baleen and sperm whales. Special opportunities exist to obtain some of the necessary information about the hearing of these cetaceans through studies of ensnared or beached animals using a Stranded Whale Auditory Test (SWAT) team, as discussed in the section below on Measurement Techniques. Although it is possible that the beached animal may be ill or damaged, the auditory system may still be normal.

Strandings, beachings, or entrapment of large cetaceans may be due to many different causes (Geraci, 1978; Klinowska, 1986; Geraci and Lounsbury, 1993) in addition to illness. Such causes as shipping vessel strikes, fishing gear entrapments, ice entrapments, isolation or orphaning of individuals, and numerous other factors that lead to a whale being marooned need not have any major effect on the animal's auditory system. Thus, it may be possible to obtain valuable data on hearing capabilities of cetaceans in these circumstances, and often this may be the only way to obtain such data.

Measurements of hearing sensitivity, temporary threshold shift, or masked threshold obtained on one individual animal characterize only that animal and may not be good estimates for the species as a whole (see subsection on Replication of Data, below). Because the animal may have a temporary or permanent hearing deficit, it would be necessary to study several individuals of each species of interest. Regarding beached or ensnared animals, there is no reason to believe that stress, injury, or disease will significantly affect the auditory system in the majority of individuals (unless, of course, the stranding is related to a hearing defect).

Auditory evoked potentials (AEP) can provide objective information about the peripheral auditory system (Bullock and Ridgway, 1972; Davis and Hirsh, 1976; Elberling and Don, 1987). Many features of the AEP, and especially the short-latency waves, are reasonably consistent across species (Allen and Starr, 1978) including cetaceans (Ridgway et al., 1981). Some components are unaffected by level of consciousness and do not require a behavioral response (Picton et al., 1974; Ridgway et al., 1981). AEPs have been used in humans to assess hearing in sleeping infants (Calloway, 1975; National Research Council, 1987) and to determine brain damage or brain death (Starr and Achor, 1975; Anon, 1968). AEP auditory information on just a few whales may provide information of critical importance in assessing potential effects of human-made low-frequency sound on mammals that are impractical to study under controlled conditions.

Dozens of individuals of several species of large whales become ensnared in fishing gear each year. For example, as many as 65 large baleen whales are ensnared off Newfoundland each June/July season (Beamish, 1973; Lien, in press). Several hours are required for people to release these animals. During the release process, it may be possible to equip the whales with electrophysiological monitoring apparatus and to obtain auditory measurements using evoked-response procedures. Major difficulties with all such work are the background noise present (here due to the release process) and adequate control of the stimulus. Awbrey et al. (1988) and Johnson et al. (1989) have shown that adequate stimuli can be presented to the whale with speakers in air and background noise can be monitored and masked. Given the current dearth of available information, and the great difficulty in working with large whales, valuable data should be obtainable from such studies.

In addition, live whales sometimes beach themselves or become isolated in tidal basins or on ice (Geraci, 1978; Lien and Stenson, 1989). On occasion, neonates or other young, apparently healthy, marine mammals of species not normally held in captivity are brought to tanks for temporary care.

Replication of Data

Aims: To determine audiometric data on multiple animals in order to understand intraspecific variance in hearing capabilities.

Hearing capabilities of terrestrial mammals generally vary among individuals of a single species, and as a consequence, measurements are generally made on several individuals in order to determine the mean and variance for a species. There is no reason to expect marine mammals to differ in this regard. Data on several individuals are necessary before it will be possible to ascertain whether, for example, measurements made after acoustic trauma are a consequence of that stimulation or whether they simply reflect normal variance within a species.

Although data are available for several different measures of underwater hearing by odontocete cetaceans and pinnipeds (Terhune and Ronald, 1972, 1975; Thomas et al., 1988, 1990; Moore and Schusterman, 1987), most studies have dealt with only one or two individual animals (but see Johnson [1966] and Seeley et al. [1976] for *Tursiops*; White et al. [1978], Awbrey et al. [1988], and Johnson et al. [1989] for white whale; and Ridgway and Joyce [1975] for the gray seal). Thus, few data exist on intraspecific variability in hearing capabilities of marine mammals. The need for data from multiple specimens is further supported by the results from two frequency discrimination studies that used different individual *Tursiops* (Jacobs, 1972; Thompson and Herman, 1975). The results from these two studies are quite different, possibly reflecting intraspecific variability in hearing capabilities in the species.

Temporary Threshold Shift

Aim: Determine sound-pressure levels that produce temporary and permanent hearing loss in marine mammals.

At this time, essentially nothing is known about the auditory aftereffects of exposure to intense sound in marine mammals, fish, or invertebrates. In land mammals, short exposures to intense sounds (greater than about 90 dB re 20 μPa —air standard) can produce a temporary hearing loss that recovers to normal within minutes, hours, or days, depending on the magnitude of the exposure and on the individual.

After repeated exposures, however, this temporary threshold shift (TTS) can develop into a permanent hearing loss (permanent threshold shift or PTS). The primary physiological site of both the TTS and PTS is the cochlear outer hair cells.

In humans and many other terrestrial mammals, the combinations of sound-pressure level and duration of exposure that produce TTS and PTS are well known. However, as far as this committee is aware, only one TTS experiment has ever been reported in fish (Popper and Clarke, 1976), and no data have been published for marine mammals. Given the enormous importance to many marine mammals of perceiving sound, a temporary or permanent loss of hearing sensitivity could potentially disrupt or shorten an animal's life.

To know whether and to what degree hearing sensitivity can be altered by human-made or natural ocean sounds of various frequencies and intensities, it is necessary to test the species of interest. As noted earlier in this chapter, the best possibilities for acquiring knowledge on some species may be research using electrophysiological measures on ensnared or beached animals. Of particular interest is whether marine mammals show a one-half-octave shift in the frequency of greatest TTS following a tonal exposure. If so, this would be strong evidence for fundamental similarities in cochlear micromechanics in marine and land mammals, a similarity that can presently only be inferred from similarities in structure (see the following subsection). Of greatest value to a full understanding of acoustic trauma in these species would be exposure experiments done in conjunction with histological studies. This approach might be feasible, notwithstanding the limitations imposed by protective legislation, if done on beached animals that cannot be returned to the water.

The issue of exposure-induced hearing loss is related to the issue of habituation (see the subsection on Habituation in the previous section). Hearing loss induced by exposure to intense sound is painless, so the creation of an exposure-induced loss does not produce a concomitant motivation for the exposed animal to avoid that high-level sound in the future. Thus, were there behavioral habituation to intense sounds, animals might, to their detriment, re-enter regions having dangerously high sound levels, thereby risking additional hearing loss. Presumably, this habituation problem would be most severe in geographical regions associated with feeding, mating, and other strong positive motivations.

Without some knowledge of the processes of TTS and PTS in marine mammals, it is impossible to predict what the long-term auditory and extra-auditory consequences might be from regular exposures to intense, low-frequency sounds.

Basic Studies of the Anatomy and Physiology of the Auditory System

Aims: To determine morphology and sound conduction paths of the auditory system in various marine mammals.

Very little is known about hearing mechanisms in marine mammals (see Popper, 1980; Ketten, 1991). Although it is known that the overall structure of the middle and inner ears is similar to that of terrestrial mammals, there are only very limited data on the detailed structure of the cochlear and more peripheral auditory structures.

One set of questions concerns how sound actually gets to the inner ear, and whether the pathways differ for different frequencies and/or in various marine mammal groups. The mechanism of stimulation of the inner ear is still controversial, and it is not clear whether the ear is stimulated by normal conduction mechanisms (via the middle ear), by bone conduction, or by some other method. Future studies need to be directed at these and other basic questions on the function of the peripheral auditory system in marine mammals, including basic analysis of the structure of the auditory apparatus in a wide variety of marine mammals. The ideal way to answer these questions would be through studies on captive specimens, but this has not been feasible with large cetaceans. Thus, studies on larger species, and especially mysticetes, might use the SWAT team approach (see subsection SWAT Team in the section on Measurement Techniques) whereby investigators, using an already prepared setup, go to the sites of strandings and record acoustic responses from entrapped animals with the specific purpose of trying to ascertain the function of specific parts of the auditory system (see subsection on Ensnared or Beached Marine Mammals, above in this section).

Clearly, the paucity of data on the auditory system of any large marine mammal is so severe that data obtained using ensnared or beached animals would provide at least an initial estimate of these animals' hearing capabilities. Through repeated studies of this type, it should be possible to get a close approximation of hearing range and sensitivity of several different species.

In addition, it may be possible to obtain valuable data in this way that correlate hearing loss in marine mammals with ear pathology. In particular, stranded animals that die should be used as a source of ear tissue for use in anatomical studies. Analysis of ear anatomy from these specimens would help to determine whether the hearing ability of a specimen was normal. If the tissue suggested ear pathology, it should be ultimately possible to correlate various hearing deficits with various pathologies. Of course, this would require a sufficient "library" of data on hearing sensitivity in a species and data on ear pathology of the same animals.

EFFECTS OF LOW-FREQUENCY SOUNDS ON THE FOOD CHAIN

Aim: To determine whether low-frequency sounds affect the behavior and physiology of organisms that serve as part of the food chain for marine mammals.

Rationale: Almost all marine mammals are predators feeding on fishes, squids, crustaceans, and other animals, and their continued survival depends on the survival and abundance of this prey. Many prey species may themselves be affected by intense low-frequency sounds, and, depending on the extent to which their availability is altered, there could be negative consequences for marine mammals as well.

Sound is important for normal behavior of many species of bony and cartilaginous fishes (reviewed in Demski et al., 1973; Myrberg, 1981). Fish use sound for a variety of reasons, including but not limited to prey detection, intraspecific communication, maintenance of schools, and predator avoidance. The sensory receptors of fish that are found in the inner ear and lateral line organs—the systems involved with detection of acoustic and hydrodynamic signals—employ the same type of sensory hair cell as found in the mammalian ear (reviewed in Popper and Fay, 1993; Popper and Platt, 1993). These cells are potentially subject to the same types of damage from exposure to intense sound as hair cells of the mammalian ear (e.g., Yan et al. 1991). Since the ear and lateral line organs are extremely important for the normal life of fish, damage to these systems would severely affect their ability to survive and reproduce, thereby affecting the food supply of marine mammals (see Popper and Platt, 1993).

Although only a few data are available on the use of sound by most marine invertebrates, including prey of marine mammals (e.g., Budelmann, 1992), some groups (e.g., cephalopods) do have highly evolved statocysts and lateral line-like systems, features which have many structural and functional similarities to the vertebrate ear and lateral line (e.g., Budelmann and Bleckmann, 1988). As in fish, damage to the major sensory receptors of these species could harm a major food source for many cetaceans.

Beyond damage to the receptors, intense sounds may alter the behavior of fish and invertebrates by affecting their ability to detect behaviorally relevant signals. For example, there is evidence that many species of deep-sea lantern fish, the myctophids, may have good hearing (Popper, 1977) and may use sound for communication (Marshall, 1967). These species make up a significant portion of the diet of a number of pinnipeds and cetaceans as well as many other fishes that are also marine mammal prey, as well as being commercially valuable (e.g., salmon, cod, hake, rockfishes, tunas) (Nafpaktitis et al., 1977). Indeed, Fitch and Brownell (1968) concluded that "if the day ever arrives that man finds it economically feasible to harvest fishes from the scattering layer (often composed predominantly of myctophids), uncontrolled exploitation could have a disastrous effect on our dolphin, porpoise, and whale populations."

In addition, high-intensity sounds may result in damage to other organ systems of these prey animals. There is laboratory evidence that such sounds can affect egg viability and growth rates of fish and invertebrates (Banner and Hyatt, 1973; Kostyuchenko, 1973; Lagardere, 1982). Thus, intense sounds may affect the availability of organisms in the food chain of marine mammals even if these organisms do not have auditory receptors.

This committee recommends that research be conducted on the ways in which fishes, squids, and crustaceans (especially krill) respond to relevant human-made sounds at levels above the ambient noise level, and on the levels that cause chronic or acute aftereffects of stimulation.

DEVELOPMENT AND APPLICATION OF MEASUREMENT TECHNIQUES

Aims: To develop tools that can enhance observation and data gathering regarding marine mammal behavior or that can protect the animals from intense human-made sounds.

Rationale: One of the major problems in understanding the behavior of marine mammals, as well as in determining the effects of human-made sounds on their behavior, is the difficulty of observing their behavior in the open ocean. A number of tools are needed to enable investigators to maintain long-term observations of animals with minimal human presence. At the same time, it may be possible to develop techniques to mitigate the effects of the intense sounds by "warning" animals away from sites where such sounds are to be used.

Tag Development

Aims: To develop tags that can be used for long-term observations of marine mammals, including studies on physiological

condition, location (in three dimensions), sound exposure levels, and acoustic behavior.

Difficulties in observing marine mammals at sea have seriously limited our ability to study their normal activities and responses to human activities. Most marine mammals live below the surface of the water much of the time, and even when at the surface, the animals may not be observable if they are far off shore or moving rapidly, or if there is bad weather, darkness, or ice.

Major advances in knowledge of the behavior of pinnipeds have occurred with recent improvements in the design and application of recoverable time/depth recorders, and more recently with satellite-linked time/depth recorders, that can be placed on pinnipeds with relative ease when they come ashore. These devices would be more suitable for studying effects of low-frequency sounds if they incorporated additional sensors and, in some cases, included a provision for remote release and data recovery.

Advances in knowledge of the behavior of cetaceans have been much less dramatic because of problems with the deployment, attachment, and recovery of tags. Accordingly, the committee recommends a focused effort to develop improved deployment, attachment, and telemetry methods for tags. For many pinniped as well as cetacean applications, satellite-linked tags are needed. However, the data capacity of the existing ARGOS system is seriously limited. A replacement system, expected in 1996, will increase data transmission capabilities and allow two-way communication between the satellites and transmitters.

Optimized data compaction algorithms are needed aboard the tags. Ultimately and ideally, new satellite systems for getting data from the tags are needed to enable much higher data rates to be obtained. Tags incorporating additional sensors are a prerequisite for much of the high-priority research on effects of low-frequency sounds on marine mammals in the wild (see the first section in this chapter).

Tags are needed that would directly or indirectly provide precise three-dimensional position data (latitude, longitude, depth). It would be extremely valuable if such a tag could also measure and record the sound levels received by the animal at different times. Ideally, the same sensor could determine sound levels before, during, and after exposure to human-made noises, and could monitor vocalization behavior of the tagged animal. Other desirable parameters include measures of animal motion (e.g., heading, pitch, speed, acceleration), foraging behavior (e.g., stomach temperature, mouth opening and closing), and physiological condition (e.g., heart rate). Tags incorporating even one or two of these new capabilities would be very useful.

Integrated Undersea Surveillance System (IUSS)

Aim: To develop means of using in-place acoustic monitoring devices to study marine mammal movement and behavior on an ocean basin scale and of following individuals or groups of animals for extended periods and distances.

Marine mammals have traditionally been studied by using visual observation or remote tags. Passive acoustic localization of vocalizing animals is a relatively unexploited technique that can compensate for some of the weaknesses of the alternative approaches. The U.S. Navy is currently encouraging the use of its integrated undersea surveillance system (IUSS) for locating and triangulating points of origin of marine mammal vocalizations (see Amato, 1993; Gagnon and Clark, 1993). IUSS is an integrated system of hydrophone arrays mounted on the sea floor in many widely separated areas allowing coverage of large parts of the North Atlantic and North Pacific. The IUSS has been used to locate calling whales at apparent ranges of hundreds of kilometers.

The IUSS might be helpful in evaluating the potential effects of intense noise sources on whales distributed over large areas of ocean. Unlike vessels, which may disturb whales before ship-based observers can even detect them, monitoring of whales with IUSS does not disturb the animals. It may be possible to use IUSS to map the density of whale calls and even to track whales, under ideal conditions. This mapping could provide a synoptic view of the vocal behavior of hundreds or thousands of animals. If a sound source were activated in the center of such a map, one could evaluate several potential disturbance responses. When disturbed, whales may either cease vocalizing, change vocalizations, or move away from or toward the disturbing sound. These responses would show up on IUSS as a lower density and/or altered distribution of whale calls.

An IUSS study would help to evaluate the range over which whales may respond to noise and would also help to integrate responses of large numbers of animals. Both kinds of data would complement intensive studies of how individual whales in one place respond to the same noise source.

SWAT Team for Studying Hearing by Ensnared or Beached Animals

Aim: To develop procedures for rapid determinations of hearing capabilities (and perhaps other physiological studies) on beached or ensnared marine mammals.

Although it is possible to study some species of smaller marine mammals in captivity, it is not feasible to study larger species under such conditions using behavioral methods. Data from larger species (especially baleen whales) are particularly important for developing an understanding of the physiological effects of high-intensity, low-frequency sounds on these animals. One approach to determining some aspects of the hearing capabilities of such species would be to record physiological evoked potentials in response to presentation of acoustic signals (see subsection Measurements on Ensnared or Beached Marine Mammals in the section on the Auditory System). Physiological evoked potentials include, for example, averaged brain-wave responses such as cortical evoked potentials and averaged brainstem responses, heart rate changes, otoacoustic emissions, and galvanic skin responses.

A major potential source of animals for such studies would be those species that are ensnared in nets for short periods of time or animals that are beached. However, since ensnarement and beaching are generally unpredictable, both in time and in place, it would be valuable to have a group of investigators prepared to depart for such sites on very short notice. Therefore the committee proposes development of a SWAT (Stranded Whale Auditory Test) team of investigators having portable equipment and highly developed techniques that would enable them to go to sites of ensnarement or beaching to obtain data on the hearing capabilities of animals. These teams should also have the training, equipment, and permits needed to collect auditory structures and inner ears from those animals that do not survive the ensnarement or beaching.

Warning Signals

Aim: To investigate the possibility of protecting marine mammals from some of the adverse effects of intense, low-frequency sounds by capitalizing on any normal avoidance reactions these animals might have to certain sounds.

Some of the negative effects of intense, low-frequency sound on marine mammals could obviously be avoided if there were no mammals in the immediate vicinity of the sound at the time of its generation. This might be accomplished through the use of stimuli that particular species naturally find offensive, alarming, or noxious. If it were possible to transmit an avoidance or warning signal for a period of time prior to producing a sound that might be dangerous to the auditory systems of some of the animals in the vicinity, considerable protection might be realized for that species as well as for the individual animals involved. Numerous reports have appeared over the years of whales fleeing from sounds of killer whales or certain human activities [see reviews by Reeves (1992) and Richardson et al. (1991)]. If these reports could be confirmed, and/or additional warning sounds found for other species, it might be possible to reduce the effects of various intense sounds on marine mammals greatly. Indeed, for some species it is possible that effective warning signals could be developed over time through the process of classical conditioning. Alternatively, the potentially damaging sounds might initially be projected at a low intensity, allowing behavioral avoidance to occur before the level is increased to high intensity.

A broader application would be to develop effective warning signals to protect marine mammals from other potentially harmful activities such as entanglement with fishing gear or collision with ships.

REFERENCES

Allen, A.R., and A. Starr. 1978. Auditory brain stem potentials in monkey (*M. mulatta*) and man. Electroenceph. Clin. Neurophysiol. 45:53–63.

Amato, I. 1993. A sub surveillance network becomes a window on whales. Science 261(5121):549–550.

Anon. 1968. Report of the Ad Hoc Committee of the Harvard Medical School to Examine the Definition of Brain Death. A definition of irreversible coma. J.A.M.A. 205(6):337–340.

Awbrey, F.T., J.A. Thomas, and R.A. Kastelein. 1988. Low-frequency underwater hearing sensitivity in belugas, *Delphinapterus leucas*. J. Acoust. Soc. Am. 84:2273–2275.

Banner, A., and M. Hyatt. 1973. Effects of noise on eggs and larvae of 2 estuarine fish. Trans. Am. Fish. Soc. 102(1):134–136.

Beamish, P. 1973. Behavior and significance of entrapped baleen whales. *In*: H.E. Winn and B.L. Olla (eds.), Behavior of Marine Animals: Current Perspectives in Research. Plenum, New York. pp. 291–309.

Budelmann, B-U. 1992. Hearing in crustacea. *In*: D.B. Webster, R.R. Fay, and A.N. Popper (eds). Evolutionary Biology of Hearing. Springer-Verlag, New York. pp. 131–139.

Budelmann, B-U., and H. Bleckmann. 1988. A lateral line analog in cephalopods: Water waves generate microphonic potentials in the epidermal head lines of *Sepia officinalis* and *Lolliguncula brevis*. J. Comp. Physiol. A 164:1–5.

Bullock, T.H., and S.H. Ridgway. 1972. Evoked potentials in the central auditory system of alert porpoises to their own and artificial sounds. J. Neurobiol. 3:79–99.

Calloway, E. 1975. Brain electrical potentials and individual psychological differences. Grune & Stratton, New York. p. 29.
Clark, C.W., and J. M. Clark. 1980. Sound playback experiments with southern right whales (*Eubalaena australis*). Science 207:663–665.
Cummings, W.C., and P.O. Thompson. 1971. Underwater sounds from the blue whale, *Balaenoptera musculus*. J. Acoust. Soc. Am. 50:1193–1198.
Davis, H., and S.K. Hirsh. 1976. The audiometric utility of brain stem responses to low-frequency sounds. Audiology 15:181–195.
Demski, L., G.W. Gerald, and A.N. Popper. 1973. Central and peripheral mechanisms in teleost sound production. Am. Zool. 13:1141–1167.
Elberling, C., and M. Don. 1987. Threshold characteristics of the human auditory brainstem response . J. Acoust. Soc. Am. 81:115–121.
Fitch, J.E., and R.J. Brownell, Jr. 1968. Fish otoliths in cetacean stomachs and their importance in interpreting feeding habits. J. Fish. Res. Board Can. 25:2561–2574.
Gagnon, G.J., and C.W. Clark. 1993. The use of U.S. Navy IUSS passive sonar to monitor the movement of blue whales. Abstr. 10th Bien. Conf. Biol. Mar. Mamm. Galveston, TX, Nov. 1993:50.
Geraci, J.R. 1978. The enigma of marine mammal strandings. Oceanus 21:38–47.
Geraci, J.R., and V.J. Lounsbury. 1993. Marine Mammals Ashore: A Field Guide for Strandings. Texas A&M Sea Grant Publication. College Station, TX. 309 pp.
Jacobs, D.W. 1972. Auditory frequency discrimination in the Atlantic bottlenosed dolphin, *Tursiops truncatus* Montague: A preliminary report. J. Acoust. Soc. Am. 52:696–698.
Johnson, C.S. 1966. Auditory thresholds of the bottlenosed porpoise (*Tursiops truncatus*), China Lake: U.S. Naval Ordnance Test Station, NOTS TP 4178.
Johnson, C.S., M.W. McManus, and D. Skaar. 1989. Masked tonal hearing thresholds in the beluga whale . J. Acoust. Soc. Am. 85:2651–2654.
Ketten, D.R. 1991. The marine mammal ear: specialization for aquatic audition and echolocation. *In*: D.B. Webster, R.R. Fay, and A.N. Popper (eds.), Evolutionary Biology of Hearing. Springer-Verlag, Berlin. pp. 717–750.
Klinowska, M. 1986. The cetacean magnetic sense—evidence from strandings. *In*: M.M. Bryden and R.J. Harrison, (eds.), Research on Dolphins. Oxford University Press, Oxford, UK. pp. 401–432.
Kostyuchenko, L.P. 1973. Effect of elastic waves generated in marine seismic prospecting on fish eggs in the Black Sea. Hydrobiol. J. 9(5):45–48.
Lagardere, J.P. 1982. Effects of noise on growth and reproduction of crangon-crangon in rearing tanks. Marine Biol. 71(2):177–186.
Lien, J. In press. Entrapments of large cetaceans in passive inshore fishing gear in Newfoundland and Labrador (1979–1990). Rep. Int. Whal. Comm., Special Issue.
Lien, J., and G.B. Stenson. 1989. Ice entrapments of blue whales (*Balaenoptera musculus*) on Newfoundland's SW coast. Canadian Atlantic Fisheries Scientific Advisory Council, working paper 01/89.
Malme, C.I., P.R. Miles, C.W. Clark, P. Tyack, and J.E. Bird. 1983. Investigations of the potential effects of underwater noise from petroleum industry activities on migrating gray whale behavior. BBN Rep. 5366. Rep. from Bolt Beranek & Newman Inc., Cambridge, MA, for U.S. Minerals Manage. Serv., Anchorage, AK. Var. pag. NTIS PB86-174174.
Malme, C.I., P.R. Miles, C.W. Clark, P. Tyack, and J.E. Bird. 1984. Investigations of the potential effects of underwater noise from petroleum industry activities on migrating gray whale behavior/Phase II: January 1984 migration. BBN Rep. 5586. Rep. from Bolt Beranek & Newman Inc., Cambridge, MA, for U.S. Minerals Manage. Serv., Anchorage, AK. Var. pag. NTIS PB86-218377.

Marshall, N.B. 1967. Sound-producing mechanisms and the biology of deep-sea fishes. *In*: W.N. Tavolga, (ed), Marine Bio-Acoustics II. Pergamon Press, Oxford. pp. 123–133.

Moore, P.W.B., and R.J. Schusterman. 1987. Audiometric assessment of northern fur seals, *Callorhinus ursinus*. Mar. Mamm. Sci. 3(1):31–53.

Myrberg, A.A., Jr. 1981. Sound communication and perception in fishes. *In*: W.N. Tavolga, A.N. Popper, and R.R. Fay (eds.), Hearing and Sound Communication in Fishes Springer-Verlag, New York. pp. 395–426.

Nafpaktitis, B.G., R.H. Backus, J.E. Craddock, R.L. Haedrich, B.H. Robison, and C. Karnella. 1977. Family Myctophidae. *In*: R.J. Gibbs Jr. (ed.), Fishes of the Western North Atlantic. Sears Foundation of Marine Research, Yale University, New Haven, CT. pp. 13–265.

National Research Council. 1987. Brainstem Audiometry of Infants. ASHA (American Speech Language-Hearing Association), January:47–55.

Picton, T.W., S.A. Hillyard, H.I. Krausz, and R. Galambos. 1974. Human auditory evoked potentials. I: Evaluation of components. Electroenceph. Clin. Neurophysiol. 36:179–190.

Popov, V.V., T.F. Ladygina, and A.Y. Supin. 1983. Electrical reactions of the delphinid (porpoise) auditory cortex as a function of time parameters of acoustic signals. Dokl. Akad. Nauk SSSR 1271:758–761.

Popper, A.N. 1977. A scanning electron microscopic study of the sacculus and lagena in the ears of fifteen species of teleost fishes. J. Morph. 153:397–418.

Popper, A.N. 1980. Sound emission and detection by delphinids. *In*: L.M. Herman (ed.). Cetacean Behavior: Mechanisms and Processes. J. Wiley and Sons, New York. pp. 1–51.

Popper, A.N., and N.L. Clarke. 1976. The auditory system of the goldfish (*Carassius auratus*): Effects of intense acoustic stimulation. Comp. Biochem. Physiol. 53A:11–18.

Popper, A.N., and R.R. Fay. 1993. Sound detection and processing by fish: Critical review and major research questions. Brain Behav. Evol. 41:14–38.

Popper, A.N. and C. Platt. 1993. Inner ear and lateral line of bony fishes. *In*: D.H. Evans (ed.), The Physiology of Fishes. CRC Press, Boca Raton, FL. pp. 99–136

Reeves, R.R. 1992. Whale responses to anthropogenic sounds: A literature review. Sci. & Res. Ser. 47. N.Z. Dep. Conserv., Wellington, New Zealand. 47 pp.

Richardson, W.J., B. Würsig, and C.R. Greene, Jr. 1986. Reactions of bowhead whales, *Balaena mysticetus*, to seismic exploration in the Canadian Beaufort Sea. J. Acoust. Soc. Am. 79 (4):1117–1128.

Richardson, W.J., C.R. Greene, Jr., C.I. Malme, and D.H. Thomson. 1991. Effects of noise on marine mammals. OCS Study MMS 90-0093; LGL Rep. TA834-1. Rep. from LGL Ecol. Res. Assoc. Inc., Bryan, TX, for U.S. Minerals Manage. Serv., Atlantic OCS Reg., Herndon, VA. 462 pp. NTIS PB91-168914.

Ridgway, S.H., and P.L. Joyce. 1975. Studies on seal brain by radiotelemetry. Rapp. P.-V. Réun. Cons. Int. Explor. Mer 169:81–91.

Ridgway, S.H., T.H. Bullock, D.A. Carder, R.L. Seeley, D. Woods, and R. Galambos. 1981. Auditory brainstem response in dolphins. Proc. Nat. Acad. Sci. USA 78:1943–1947.

Seeley, R.L., W.F. Flanigan, Jr., and S.H. Ridgway. 1976. A technique for rapidly assessing the hearing of the bottlenosed porpoise, *Tursiops truncatus*. NUC-TP-522. Naval Undersea Center, San Diego, CA. 15 pp. NTIS AD-A029178.

Starr, A., and L.J. Achor. 1975. Auditory brain stem responses in neurological disease. Arch. Neurol. (Chic.) 32:761–768.

Terhune, J.M., and K. Ronald. 1972. The harp seal, *Pagophilus groenlandicus* (Erxleben, 1777) III. The underwater audiogram. Can. J. Zool. 50:565–569.

Terhune, J.M., and K. Ronald. 1975. Underwater hearing sensitivity of two ringed seals (*Pusa hispida*). Can. J. Zool. 53:227–231.

Thomas, J., N. Chun, W. Au, and K. Pugh. 1988. Underwater audiogram of a false killer whale (*Pseudorca crassidens*). J. Acoust. Soc. Am. 84:936–940.

Thomas, J., P. Moore, R. Withrow, and M. Stoermer. 1990. Underwater audiogram of a Hawaiian monk seal (*Monachus schauinslandi*). J. Acoust. Soc. Am. 87:417–420.

Thompson, P.O., W.C. Cummings, and S.J. Ha. 1986. Sounds, source levels, and associated behavior of humpback whales, Southeast Alaska. J. Acoust. Soc. Am. 80:735–740.

Thompson, R.K.R., and L.M. Herman. 1975. Underwater frequency discrimination in the bottlenosed dolphin (1–140kHz), and the human (18 kHz). J. Acoust. Soc. Am. 57:943–948.

White, M.J., Jr., J. Norris, D. Ljungblad, K. Baron, and G. di Sciara. 1978. Auditory thresholds of two beluga whales (*Delphinapterus leucas*) . HSWRI Tech. Rep. 78–109. Rep. from Hubbs/Sea World Res. Inst., San Diego, CA, for Naval Ocean Systems Center, San Diego, CA. 35 pp.

Yan, H.Y., W.M. Saidel, J. Chang, J.C. Presson, and A.N. Popper. 1991. Sensory hair cells of the fish ear: Evidence of multiple types based on ototoxicity sensitivity. Proc. Roy. Soc. Ser. B 245:133–138.

Appendix A

Comparison of Sound-Pressure Reference Levels in Air and Water

Because of its enormous range, sound amplitude is often described in logarithmic units, decibels (dB). Some small pressure is used as a reference pressure, and any other sound pressure is described as a level above that reference pressure. The reference level is analogous to sea level in our specification of the height of some land mass.

Unfortunately, researchers studying sound in water and air typically use a different reference pressure. In water, the common reference has been 1 micropascal (1 μPa, or one millionth of a pascal; a pascal is 1 newton per square meter). In air, the common reference is 20 μPa, because that is near the absolute threshold for a normal human listener for a sound frequency of 1,000 Hz. This level is called sound-pressure level (SPL). These two references are, therefore, 26 dB apart (20 log 20 = 26). Table A-1 gives both levels for some airborne sounds, and also shows the levels of some marine mammal sounds under water.

The difference in reference pressure level is one complication in comparing sound in air with sound in water. Another is that, because the impedances of air and water differ, the actual power flow in them differs even if the pressures are the same. For example, a spherical sound source radiating a pressure of 1 dyne per square centimeter in air generates about 2.5×10^{-9} watts per square centimeter. The same source in water radiating the same pressure generates about 4.7×10^{-13} W/cm^2—an intensity ratio of about 5,000. Thus, great care must be taken in comparing sound levels in air with sound levels in water.

TABLE A-1 Typical Airborne Sounds and Some Sound Levels of Marine Mammals

Typical sound in air/ *marine mammal sound*	Water standard (dB re 1 μPa)	Air standard (dB re 20 μPa SPL)
Threshold of human hearing (1,000 Hz)	[26]	0
Very quiet living room	[66]	40
Seal threshold underwater (1,000 Hz)	80	[54]
Normal speech (1 meter)	[86]	60
Beluga threshold (1,000 Hz)	100	[74]
Lion's roar (10 meters)	[116]	90
Jet airliner (10 meters)	[130]	104
Fin whale call (100 meters)	140	[114]
Human threshold of pain (at ear drum)	[166]	140
Some military artillery	[186]	160
Beluga echolocation call (1 meter)	220	[194]

Source: Adapted from Kryter (1985) and Richardson et al. (1991).
NOTE: Bracketed levels are nominal levels after conversion to alternate medium.

References

Kryter, K.D. 1985. The effects of noise on man, 2nd ed. Academic Press, Orlando, FL. 688 pp.

Richardson, W.J., C.R. Greene, Jr., C.I. Malme, and D.H. Thomson. 1991. Effects of noise on marine mammals. OCS Study MMS 90-0093; LGL Rep. TA834-1. Rep. from LGL Ecol. Res. Assoc. Inc., Bryan, TX, for U.S. Minerals Manage. Serv., Atlantic OCS Reg., Herndon, VA. 462 pp. NTIS PB91-168914.

Appendix B

An Introduction to Marine Mammals

Three orders of the class Mammalia contain marine mammals: Cetacea, Carnivora, and Sirenia. (There is a section on each of the three orders in this appendix. Each includes a table listing the marine mammals in that order.) The size range of marine mammals is immense, varying from a newborn sea otter weighing perhaps 1 kilogram (kg) to the largest female blue whale weighing about 100,000 kg. Their habitats are also quite varied, encompassing all seas and numerous coastal areas and shores as well as some freshwater lakes and rivers.

All extant Cetacea and Sirenia normally spend their entire lives in water. In contrast, marine mammals of the order Carnivora are semiaquatic, often hauling out on land. Some of these semiaquatic mammals spend considerable periods of time (many months) in the water, often hundreds or even thousands of kilometers at sea away from haul-out or breeding areas. Families of marine mammals of the order Carnivora are these: Otariidae, or eared seals (fur seals and sea lions); Odobenidae (walrus); Phocidae (true seals); Mustelidae (chungungo and sea otter); and Ursidae (polar bear).

In the United States, all marine mammals are legally protected by the Marine Mammal Protection Act of 1972 (MMPA). The species of the order Cetacea and suborder Pinnipedia that live strictly in freshwater are also protected under this law, and thus they are included in this discussion. Under the 1972 act, the U.S. Fish and Wildlife Service (FWS) of the U.S. Department of the Interior is responsible for administering regulations concerning polar bears, walrus, sirenians, and sea otters; the National Marine Fisheries Service of the U.S. Department of Commerce is responsible for whales, dolphins, seals, sea lions, and other marine mammals not regulated by FWS. In addition, another independent body established under the MMPA, the Marine Mammal Commission, maintains a scientific committee to advise on issues related to marine mammal conservation. Those species designated as endangered are further protected under the Endangered Species Act of 1973.

Threats to marine mammal populations worldwide are many. They result, for example, from fishing (the use of gillnets, driftnets, ghost nets, long lines, the yellowfin tuna purse seine, rolling hooks), pollution (agricultural runoff, industrial waste, petroleum spills, trash dumping), deforestation and development of the rain forests, damming, oil field development, mining, heavy-vessel traffic, and other human activities.

Twenty species of marine mammals are listed as endangered under U.S. provisions (see the tables in this appendix), although some of them appear to be gaining in population and may be removed from endangered status (Brownell et al., 1989). For example, the eastern or California stock of the gray whale *Eschrichtius robustus* has apparently recovered from severe exploitation and thus was proposed for removal from the endangered species list by the U.S. Department of Commerce as of 7 January 1993 (Marine Mammal Commission, 1992). Some other marine mammals, such as the Gulf of California porpoise (vaquita) and the Yangtze River dolphin (baiji), appear to be headed for extinction. However, the majority of species are not endangered or seriously threatened.

CETACEA: WHALES, DOLPHINS, PORPOISES

The larger cetaceans include Physeteridae (sperm whales), Ziphoidea (beaked whales and bottlenose whales), and suborder Mysticeti (baleen whales) (Table B-1). Eight of the largest species are listed as endangered.

The sperm whale has been on the U.S. List of Endangered and Threatened Wildlife, even though there may be more than a million sperm whales in the world's oceans. Some specific populations are apparently depleted even though the world population is relatively large.

Because of their past exploitation by whalers and recent publicity about this exploitation, most large baleen whales are assumed by the public to be endangered species. However, in the absence of the

TABLE B-1 Classification of Living Mammals of the Order Cetacea

Suborder, family, genus, species	Common name
I. Suborder Odontoceti	
Superfamily Platanistoidea	
Family Platanistidae	
Platanista gangetica	Ganges River dolphin, Ganges susu
Platanista minor *	Indus River dolphin, Indus susu
Family Pontoporiidae	
Subfamily Lipotinae	
Lipotes vexillifer *†	baiji, Yangtze, or Chinese River dolphin
Subfamily Pontoporiinae	
Pontoporia blainvillei	franciscana, cachimbo, La Plata dolphin
Family Iniidae	
Inia geoffrensis †	boto, boutu, bufeo, Amazon River dolphin
Superfamily Delphinoidea	
Family Monodontidae	
Subfamily Orcaellinae	
Orcaella brevirostris	Irrawaddy dolphin, pesut
Subfamily Delphinapterinae	
Delphinapterus leucas †	white whale, beluga
Subfamily Monodontinae	
Monodon monoceros	narwhal
Family Phocoenidae	
Subfamily Phocoeninae	
Phocoena phocoena†	harbor porpoise
Phocoena spinipinnis	Burmeister's porpoise
Phocoena sinus *	vaquita, Gulf of California harbor porpoise
Neophocaena phocaenoides	finless porpoise
Subfamily Phocoenoidinae	
Australophocaena dioptrica	spectacled porpoise
Phocoenoides dalli	Dall's porpoise
Family Delphinidae	
Subfamily Steninae	
Steno bredanensis	rough-toothed dolphin
Sousa chinensis	Indopacific hump-backed dolphin
Sousa plumbea	plumbeous dolphin
Sousa teuszii	Atlantic hump-backed dolphin
Sotalia fluviatilis	tucuxi
Subfamily Delphininae	
Lagenorhynchus albirostris	white-beaked dolphin
Lagenorhynchus acutus	Atlantic white-sided dolphin
Lagenorhynchus obscurus	dusky dolphin
Lagenorhynchus obliquidens	Pacific white-sided dolphin
Lagenorhynchus cruciger	hourglass dolphin
Lagenorhynchus australis	Peale's dolphin

Suborder, family, genus, species	Common name
Grampus griseus	Risso's dolphin
Tursiops truncatus	bottlenose dolphin
Stenella frontalis	Atlantic spotted dolphin
Stenella attenuata	pantropical spotted dolphin
Stenella longirostris	spinner dolphin
Stenella clymene	clymene dolphin
Stenella coeruleoalba	striped dolphin
Delphinus delphis	common dolphin
Lagenodelphis hosei	Fraser's dolphin
Subfamily Lissodelphinae	
Lissodelphis borealis	northern right whale dolphin
Lissodelphis peronii	southern right whale dolphin
Subfamily Cephalorhynchinae	
Cephalorhynchus commersonii	Commerson's dolphin
Cephalorhynchus eutropia	black dolphin, Chilean dolphin
Cephalorhynchus heavisidii	Heaviside's dolphin
Cephalorhynchus hectori	Hector's dolphin
Subfamily Globicephalinae	
Peponocephala electra	melon-headed whale, electra dolphin
Feresa attenuata	pygmy killer whale
Pseudorca crassidens †	false killer whale
Orcinus orca †	killer whale
Globicephala melas	long-finned pilot whale
Globicephala macrorhynchus	short-finned pilot whale
Superfamily Ziphoidea	
Family Ziphiidae	
Tasmacetus shepherdi	Shepherd's beaked whale
Berardius bairdii	Baird's beaked whale
Berardius arnuxii	Arnoux's beaked whale
Mesoplodon pacificus	Longman's beaked whale
Mesoplodon bidens	Sowerby's beaked whale
Mesoplodon densirostris	Blainville's beaked whale
Mesoplodon europaeus	Gervais' beaked whale
Mesoplodon layardii	strap-toothed whale
Mesoplodon hectori	Hector's beaked whale
Mesoplodon grayi	Gray's beaked whale
Mesoplodon stejnegeri	Stejneger's beaked whale
Mesoplodon bowdoini	Andrew's beaked whale
Mesoplodon mirus	True's beaked whale
Mesoplodon ginkgodens	ginkgo-toothed beaked whale
Mesoplodon carlhubbsi	Hubb's beaked whale
Ziphius cavirostris	Cuvier's beaked whale
Hyperoodon ampullatus	northern bottlenose whale
Hyperoodon planifrons	southern bottlenose whale

Suborder, family, genus, species	Common name
Superfamily Physeteroidea	
Family Physeteridae	
Subfamily Physeterinae	
Physeter macrocephalus *	sperm whale
Family Kogiidae	
Kogia breviceps	pygmy sperm whale
Kogia simus	dwarf sperm whale
II. Suborder Mysticeti	
Family Balaenidae	
Balaena mysticetus *	bowhead whale
Eubalaena australis *	southern right whale
Eubalaena glacialis *	northern right whale
Family Neobalaenidae	
Caperea marginata	pygmy right whale
Family Eschrichtiidae	
*Eschrichtius robustus**[1]	gray whale
Family Balaenopteridae	
Subfamily Balaenopterinae	
Balaenoptera acutorostrata	minke whale
Balaenoptera borealis *	sei whale
Balaenoptera edeni	Bryde's whale
Balaenoptera musculus *	blue whale
Balaenoptera physalus *	fin whale, finback
Subfamily Megapterinae	
Megaptera novaeangliae *	humpback whale

NOTE: * = endangered species; † = species for which some audiometric inf,ormation has been published. [1]The California stock of gray whales has been recommended for delisting by NMFS.

resumption of full commercial whaling, most of these larger cetaceans, including some species of baleen whales that appear on the Endangered Species List are, as species, in reality not endangered. Some of these species have been completely protected for many years, and all are currently protected by the moratorium on commercial whaling promulgated by the International Whaling Commission. As with the sperm whales, however, some populations of large baleen whales remain depleted. Brownell et al. (1989) suggest that some of these large whales be removed from the Endangered Species List and that some small cetaceans be added. With the possible exception of the northern right whale, none of the large cetacean species is currently in peril of extinction (Perrin, 1988).

There are more than 40 species of smaller cetaceans, dolphins and porpoises, found worldwide. While no cetacean species has been driven to extinction by human endeavors (Perrin, 1988), four species of the smaller cetaceans are in jeopardy in the coming decades if certain human activities in their habitats are not changed (Brownell, 1991; Norris, 1992). These include the baiji, *Lipotes vexillifer*; the Chilean dolphin, *Cephalorhynchus eutropia*; the Indus River dolphin, *Platanista minor*; and the vaquita, *Phocoena sinus*, of the Gulf of California.

THE CARNIVORA: PINNIPEDS, SEA OTTERS, POLAR BEARS

The marine Carnivora all spend some time on land or sea ice to breed and bear their young. The breeding areas are therefore especially sensitive to human encroachment. A few species live in freshwater. Marine mammals in the order Carnivora include fur seals and sea lions, true seals, walrus, chungungo and sea otters, and polar bears (Table B-2).

Pinnipedia

Pinniped means "feather footed," and the suborder Pinnipedia includes three families: Otariidae (fur seals and sea lions), Odobenidae (walrus), and Phocidae (true seals). Historically, almost all pinnipeds were hunted for fur, meat, oil, or ivory. Brownell (1991) suggests that, although most pinniped species will probably experience an increase in number during the 1990s, at least three species—northern fur seal, *Callorhinus ursinus*; Steller sea lion, *Eumetopias jubatus*; and Mediterranean monk seal, *Monachus monachus*—will continue to decline in the absence of stronger corrective measures. One pinniped species, the Caribbean monk seal, *Monachus tropicalis*, has apparently become extinct during this century (Kenyon, 1977).

Sea Otters

The sea otter of the North Pacific, *Enhydra lutris*, is a coastal animal that is often associated with kelp beds just off the coasts of California, British Columbia, and Alaska.

The marine otter off the coasts of Peru and Chile, chungungo, *Lutra felina*, is severely endangered because the animals are hunted for their prized pelts on the Chilean coast and because Peruvians often shoot them as a menace to fishing (Miller and Rottmann, 1983). Another large otter, *Lutra longicaudis* of northeast Brazil, is threatened by fishing, clandestine hunting, and habitat degradation (Almeida et al., 1991).

TABLE B-2 Marine Mammals (Pinnipeds, Otters, Polar Bears) of the Order Carnivora

Family, genus, species	Common name
Family Otariidae of suborder Pinnipedia	
Eumetopias jubatus	Steller sea lion, northern sea lion
Zalophus californianus †	California sea lion
Otaria flavescens	South American sea lion
Neophoca cinerea	Australian sea lion
Phocarctos hookeri	New Zealand sea lion
Callorhinus ursinus †	northern fur seal
Arctocephalus townsendi *	Guadalupe fur seal
Arctocephalus philippii *	Juan Fernández fur seal
Arctocephalus galapagoensis	Galápalgos fur seal
Arctocephalus australis	South American fur seal
Arctocephalus pusillus	Cape fur seal, South African fur seal, Tasmanian fur seal, Victorian fur seal
Arctocephalus forsteri	New Zealand fur seal, West Australian fur seal
Arctocephalus gazella	Kerguelen fur seal, Antarctic fur seal
Arctocephalus tropicalis	Amsterdam Island fur seal
Family Odobenidae of suborder Pinnipedia	
Odobenus rosmarus	walrus
Family Phocidae of suborder Pinnipedia	
Phoca vitulina †	harbor seal
Phoca largha	larga seal, spotted seal
Phoca hispida †	ringed seal
Phoca sibirica	Caspian seal
Phoca groenlandica †	harp seal, Greenland seal
Phoca fasciata	ribbon seal
Erignathus barbatus	bearded seal
Cystophora cristata	hooded seal, bladdernose seal
Halichoerus grypus †	gray seal
Monachus monachus *	Mediterranean monk seal
Monachus tropicalis ‡	West Indian monk seal, Caribbean monk seal
Monachus schauinslandi *†	Hawaiian monk seal
Mirounga leonina	southern elephant seal
Mirounga angustirostris	northern elephant seal
Lobodon carcinophagus	crabeater seal
Ommatophoca rossii	Ross seal
Hydrurga leptonyx	leopard seal
Leptonychotes weddelli	Weddell seal
Family Mustelidae	
Lutra felina *	Chungungo, marine otter, gato marino
Enhydra lutris	sea otter
Family Ursidae	
Ursus maritimus	polar bear

NOTES: The otters and bear listed in this table are regarded as marine mammals. Of course, most bears and otters are carnivores but not marine mammals.

* = endangered species; † = species for which some audiometric data are available; ‡ = extinct species.

Polar Bears

Polar bears from the family Ursidae may be seen in the water some distance at sea. However, they spend most of the year on sea ice or, in the absence of ice, on land. They prey primarily on ringed seals (*Phoca hispida*). Polar bears also kill other marine mammals such as bearded seals (*Erignathus barbatus*), white whales (*Delphinapterus leucas*), and the narwhal (*Monodon monoceros*). Ringed seal populations are large, but white whales and narwhals, although not endangered,1 are much less numerous.

THE SIRENIA: MANATEES AND DUGONGS (SEA COWS)

Although they receive much less attention by the media, the Sirenia are probably more endangered than any of the great whales with the possible exception of right whales. In fact, a sirenian species has become extinct in modern times (in the 1800s)—the great northern sea cow, or Steller's sea cow, *Hydrodamalis gigas*, of the Bering Sea (Nishiwaki and Marsh, 1985). Dugongs and manatees are the two living genera of Sirenia (Table B-3). They represent two distinct families of plant-grazing marine mammals that are found in separate parts of the world. Dugongs occur in tropical and subtropical shallows of the Indo-Pacific region—northern Australia, the Guangxi coast of China, Indonesia, the Aru Islands, Sri Lanka, India, the Persian Gulf, the Arabian Peninsula, the East African coast, and Madagascar.

Other than the dugong, the elephant and the hyrax are the closest living relatives of the manatee. There are three living species of

TABLE B-3 Living and Recent Members of the Totally Aquatic Order Sirenia (Sea Cows)

Family, genus, species	Common name
Order Sirenia (sea cows)	
Family Dugongidae	
Dugong dugon *	dugong
Hydrodamalis gigas ‡	Steller's sea cow
Family Trichechidae	
Trichechus manatus * †	Caribbean manatee
Trichechus inunguis *	Amazonian manatee
Trichechus senegalensis	West African manatee

NOTES: * = endangered species; † = some auditory information is available; ‡ = extinct species.

manatees: the West Indian or Florida manatee, *Trichechus manatus*; the West African manatee, *T. senegalensis*; and the Amazonian manatee, *T. inunguis*. Manatees prefer shallow estuaries and swampy areas where aquatic plants are abundant (Caldwell and Caldwell, 1985). They are at times found in freshwater, brackish water, or marine waters.

West Indian or Caribbean manatees range from Georgia on the coast of the southern United States to the coast of Brazil; however, there is a very patchy distribution over this wide range. The Amazonian manatee is found in the Amazon basin and possibly in parts of the Orinoco River system. The West African manatee is apparently most abundant in the Niger River and its tributaries (Nishiwaki et al., 1982), although there are areas of dense manatee population along the coast of Sierra Leone.

REFERENCES

Almeida, R.T., G.P. Pimentel, and F.J.L. Silva. 1991. Occurrence of Otter *Lutra longicaudis* (Mammalia-Mustelidae) in mangrove area, Pernambuco State—NE Brazil. Abstract from: Ninth Biennial Conference on the Biology of Marine Mammals, Chicago Zoological Society, Chicago. p. 1.

Brownell, R.L., Jr. 1991. Marine Mammal Populations in the 1990's: Status, Problems and Research. IBI Reports. 2:1–10. (International Marine Biological Research Institute, Kamogawa, Japan.)

Brownell, R.L., Jr., K. Ralls, and W.F. Perrin. 1989. The plight of the 'forgotten' whales. Oceanus 32 (1):5–11.

Caldwell, D.K., and M.C. Caldwell. 1985. Manatees, *Trichechus manatus* Linnaeus, 1758; *Trichechus senegalensis* Link, 1795 and *Trichechus inunguis* (Natterer, 1883). *In*: S.H. Ridgway and R.J. Harrison (eds.), Handbook of Marine Mammals, vol. 3. The Sirenians and Baleen Whales. Academic Press, London. pp. 33–66.

Kenyon, K.W. 1977. Caribbean monk seal extinct. J. Mammal. 58:97–98.

Marine Mammal Commission. 1992. Annual Report of the Marine Mammal Commission: Calendar Year 1991. A report to Congress. U.S. Marine Mammal Commission, Washington, DC.

Miller, S.D., and J. Rottmann. 1983. Endangered mammals of Chile: Status and conservation. Biol. Conserv. 25:335–352.

Nishiwaki, M., and H. Marsh. 1985. Dugong, *Dugong dugon* (Muller, 1776). *In*: S.H. Ridgway and R.J. Harrison (eds.), Handbook of Marine Mammals, vol. 3. The Sirenians and Baleen Whales. Academic Press, London. pp. 1–31.

Nishiwaki, M., M. Yamaguchi, S. Shikota, S. Uchida, and T. Kataoka. 1982. Recent survey on the distribution of African manatee. Sci. Rep. Whales Res. Inst. 34:137–147.

Norris, K. 1992. Dolphins in Crisis. Nat. Geog. 182(3):2–35.

Perrin, W. F. 1988. Dolphins, Porpoises, and Whales: An Action Plan for the Conservation of Biological Diversity: 1988–1992. IUCN, Gland, Switzerland.

Appendix C

Comparison of Yearly Sound Energy from Oceanographic Research and Supertankers

Although the duration of the transmission in a typical oceanographic experiment varies depending on the experiment, a total broadcast time of 100 hours would be longer than most experiments. We will assume that 10 such experiments are conducted each year. The total yearly energy in a 1-Hz band produced by 10 typical oceanographic experiments would be about 66 dB higher than the average sound level [10 log (60×600×100×10) = 65.6 dB]. The sound levels generated by oceanographic experiments vary greatly. An average level is probably near 200 dB (re 1 µPa—water standard). The entire sonic energy generated by these 10 oceanographic surveys is, therefore, about 266 dB (re 1 µPa—water standard) in the frequency region near 50 Hz.

There are about 127 supertankers presently operating in the oceans of the world at any one time.[1] At its typical speed of 15 to 22 knots, the average supertanker produces a source level (calculated at 1 m from the source) having a spectrum level (energy in a 1-Hz band) of about 187 dB at 50 Hz and about 232 dB at 2 Hz (re 1 µPa—water standard) (personal communication, Joal J. Newcomb to the Committee on Low-frequency Sound and Marine Mammals, 1993). There is considerable variability in the noise levels generated by different

[1] Joal J. Newcomb, Code 7176, Naval Research Laboratory, Stennis Space Center, Mississippi 39529.

vessels, but the level just given is the average. The total yearly energy in a 1-Hz band produced by the supertanker fleet is 96 dB higher than the average for a single vessel [10 log $(60\times60\times24\times365\times127) = 96.03$ dB]. The entire sonic energy generated by the fleet of super-tankers is, therefore, about 283 dB near 50 Hz and about 328 dB near 2 Hz (dB re 1 μPa—water standard).